ISBN: 978-1-7366270-1-3

Cataloging in Publication data
Author
The Matahematical System of the Bible Chronology Leaflet /
Gregor Georg Gimmel, The Table of the Bible Chronology
with short comments

Library of Congress Control Number LCCN 2021904569

Disclaimer:
Please, do not believe anything that is written in this book. Please, do fasting, go to your knees, read the scriptures and ask God to reveal you the truth. Only The Word of God is the Truth. Do not believe what humans tell you. Only ask Jesus, ask the Holy Spirit until He Answers your Question and Shows You The Truth.

The Book of Jasher:
Salt Lake City, Published by J.H. Parry & Company
1887 not otherwise copyrighted.
Taken from sacred-texts.com,
John Bruno Hare

The Interlinear Bible (INL)
From Biblehub.com
No copyright available.

The English Translation of the Greek Septuagint Bible
Compiled from the Translation by Sir Lancelot C. L. Brenton
1851 at ecmarsh.com
Is in the public domain. No copyright available.

Credits: God

Country and printing number
Printed and bound in
First printing in 2022

Publishers name: Gregor Georg Gimmel

First paperback edition April 2022

Book design by Gregor Georg Gimmel
Map by Gregor Georg Gimmel

1. THE MATHEMATICAL SYSTEM OF THE BIBLE CHRONOLOGY

The Bible Chronology Consists of A Series of Nine Covenants Between God and Mankind

This short paper is targeted at anyone new as well as experienced in learning about the Bible. The following gives you an overview of the entire Bible History from the point of Creation to Salvation, from God Father to God Son. Nine covenants define the Bible as a complete History and the whole of the Bible Chronology. Humankind is represented in each covenant by a particular person:

#	Covenant Name	Location	Representative	Year
1	Covenant of Life and Death	In the Garden of Eden = Pre-Flood Israel	Adam	0
2	Covenant of Rapture	Location Unknown, likely near Eden	Enoch	987
3	Covenant of the Rainbow	Ararat Mountains, Eastern Turkey	Noah	1657
4	Covenant of Circumcision	Canaan/Israel/Mamre/Hebron	Abraham	2177
5	Covenant of the Law	Mount Sinai/Jabal al-Lawz/Saudi Arabia	Moses	2607
6	Covenant of the Temple I	Israel/Jerusalem	Solomon	3087
7	Covenant of the End of Temple I	Israel/Jerusalem	Nebuchadnezzar	3507
8	Covenant of the Temple II	Israel/Jerusalem	Zerubbabel	3577
9	Covenant of Eternal Life	Israel/Jerusalem	Jesus	4117

The Luke 3 Generation List

The 77 generations from God Father to God Son in Luke 3 is the framework of the Bible Chronology starting with Creation and leading up to Salvation. The Bible's hidden chronology begins in Genesis 5 (Adam to Noah) and continues in Genesis 11 (Noah to Abraham). Luke 3 provides the entire framework from Adam to Jesus. It assures THERE ARE NO GAPS in this UNSEEN BIBLE CHRONOLOGY. The number '77' of the 77 generations in Luke 3:23-38 stands for the completeness of this generation list through all ages from God Father to God Son. It provides us an accurate, complete, and gap-free hidden genealogy that maps into the Bible's chronology and calendar. There are no gaps in this generation list because '7' AND '77' stands for completeness, holiness, and godliness. During his last moment before his sacrificial death, God's Son cried out to God Father in Aramaic ELI: "Eli, Eli, lema sabachthani" (My God, My God, why have You forsaken Me?). God left His Son on Good Friday, the sixth day, to pay the price of all sins of all humanity of all ages. The day of Jesus' death was the sixth day, the before the Sabbath. The number '77' stands for the direct relationship between God Father and God Son. When this Bible Chronology is carefully built up, it tells you the exact years of the NINE SIGNIFICANT BIBLE COVENANTS. The Mathematical System of the Bible Chronology is an instrument that provides the precise years of birth, death, and rapture of Bible personalities throughout Bible History from Creation to Salvation, from Adam to Jesus.

The Mathematical System of the Bible Chronology

The Mathematical System of the Bible Chronology is a matrix of three pairs of 'multiple of ten' numbers that add up to 'multiple of a hundred' numbers which add up to 'multiple of a thousand': (420+480 → 900; 520+480 → 1000; 670+430 → 1100): (900+1000+1100 → 3000). A fourth quartet of 'multiple of ten' numbers adds up to the number 1700: (670+420+70+540 → 1700)

The Mathematical System of the Bible Chronology's grid provided the exact AD and BC numbers when most of the Bible personalities lived on earth.

The Bible Calendar ending in 7, the World Calendar ending in 6

The number seven stands for the day of rest, the Sabbath, the last day of Creation when the creator rested. Seven represents God's completion of the design. It stands for Adam and humanity's unity with God, resulting in Eternal Life. It means holiness, godliness, victory over sin, and Satan through the sacrificial death of Jesus Christ.

The number '777' resembles God's Trinity (Father, Son, and Holy Spirit). '777' compares Adam before the fall into sin. God created Adam with the Holy Spirit inside of him. When Adam fell into sin, the Holy Spirit departed. Adam became a 2/3 of a Being. 2/3 stands for the '666'. After falling into sin, Adam realized he was naked. People who repent and invite Jesus Christ into their hearts receive the Holy Spirit into their hearts as a '777' - mark of God.' This person is called a "born again" and "follower of Christ." The Holy Spirit has become that person's counselor, represented by the number '777'.

The number six stands for Adam's Creation and the day before the Sabbath (Good Friday), the day when Jesus Christ gave his life for the sins of humankind. Six represents separation from God when Jesus died for your sins resulting in Eternal Death. It is also standing for evilness, the worship of Satan, false religion, occultism, and the rejection of Jesus Christ.

'666' in Revelation 13:18 stands for eternal separation from God. It is the unholy trinity of Satan. People who receive the '666' or 'mark of the beast' will turn into hybrid human machines in the future. They will be remote-controlled by Satan's reign over the entire world. '666' will end all democracy; it will end freedom of speech and freedom of thinking, freedom of religion, and freedom of choice. The '666' marks of the beast will prevent people from realizing they did something wrong. They cannot repent and confess their sins. '666' will deter people from getting saved by the sacrificial death of Jesus. That is why God says they cannot receive Salvation. People will be persecuted and die as martyrs if they choose not to accept the beast's666' marks. However, people who repent and receive Jesus in their hearts will–in my prophetical understanding from the Bible–will escape this situation as written in the Gospel of Luke 21:36. These people are going to be "accounted worthy to escape" the 7-year tribulation time of Satan. People receiving the '666' marks of the beast will be separated eternally from God.

A Great World Deception will cause humanity to accept the '666' marks of the beast. Only small children will survive this time since they are unaware of all evilness—the Bible calls' 666' the number of

Adam. Adam was created on the sixth day. As soon as Adam and Eve bit into the fruit from the tree of good and evil knowledge, the Holy Spirit left them. They were separated from God. This is represented by the number '6'. '6' (Adam separated from God) +1 (Jesus, God, Holy Spirt) = '7' (A holy saved person, saved by the sacrificial death of Jesus, a born-again believer filled with the Holy Spirit).

Galatians 3:17 and the 'Multiple of Ten' numbers from Covenant to Covenant

Galatians 3:17 is the crucial verse that opens the door to us to understand the Bible contains a hidden covenant-to-covenant chronology based on 'multiple of ten' numbers. The number of years between two covenants is always a 'multiple of ten' number–except the very first covenant-to-covenant number '98**7**'. These are the 98**7** years between the Covenant of Life & Death with Adam and the Covenant of the Rapture of Enoch 98**7** years later–see Genesis 5: (130+105+90+70+65+162+65+300 = 987)

The number '987' after the Covenant of Life & Death and the number '586 BC'.

THE BIBLE CHRONOLOGY and BIBLE CALENDAR: Because each successive covenant is a 'multiple of ten' number, each successive Covenant contains the **number 7** in the last cipher of the year count: Adam Life & Death 0, Enoch Rapture 98**7**, Noah Rainbow = 987+670=165**7**, Abraham Circumcision = 987+670+520=217**7**, Moses Law = 260**7**, Solomon's Temple Start 987+670+520+480 = 308**7**, Nebuchadnezzar Temple End 987+670+520+480+420 = 350**7**, Zerubbabel 2nd Temple start = 987+670+520+480+420+70 = 357**7**, Jesus Life Eternal = 987+670+520+480+420+70+540 = 411**7**

THE WORLD CHRONOLOGY and WORLD CALENDAR: The only Bible event the World Calendars has historically exactly captured is NOT the crucifixion of Jesus. It was when the Babylonian King Nebuchadnezzar forces burned the 1st Temple in the year 586 BC. The year and possibility of 586 BC are calculable via Babylonian and Assyrian calendars.

The year 586 BC is a reference point from which, with the help of 'multiple of ten' numbers in the Bible Chronology's mathematical system, provide all historical dates in our Anno Domini calendar system. The multiple of ten show always year counts ending with the **number 6 for the last cipher in the year count of the World Calendar**:

58**6** BC Nebuchadnezzar Temple End, 586-70 = 51**6** BC Zerubbabel 2nd Temple start, Solomon's Temple Start 586+420 = 100**6** BC, Moses Law 586+420+480 = 148**6** BC, Abraham Circumcision 586+420+480+430 = 191**6** BC, Noah Rainbow 586+420+480+430+520 = 243**6** BC, Enoch Rapture 586+420+480+430+520+670 = 310**6** BC

EXCEPTIONS: Exceptions are Adam Death & Life 586+420+480+430+520+670+987 = **4093 BC** and Jesus Eternal Life 586-70-540+1 = **25 AD** (+1 because there is no year zero in the World Calendar). Altogether there are 4117 years from the Covenant of Life & Death with Adam and the Covenant of Eternal Life with Jesus.

2. THE BIBLE CHRONOLOGY MATRIX

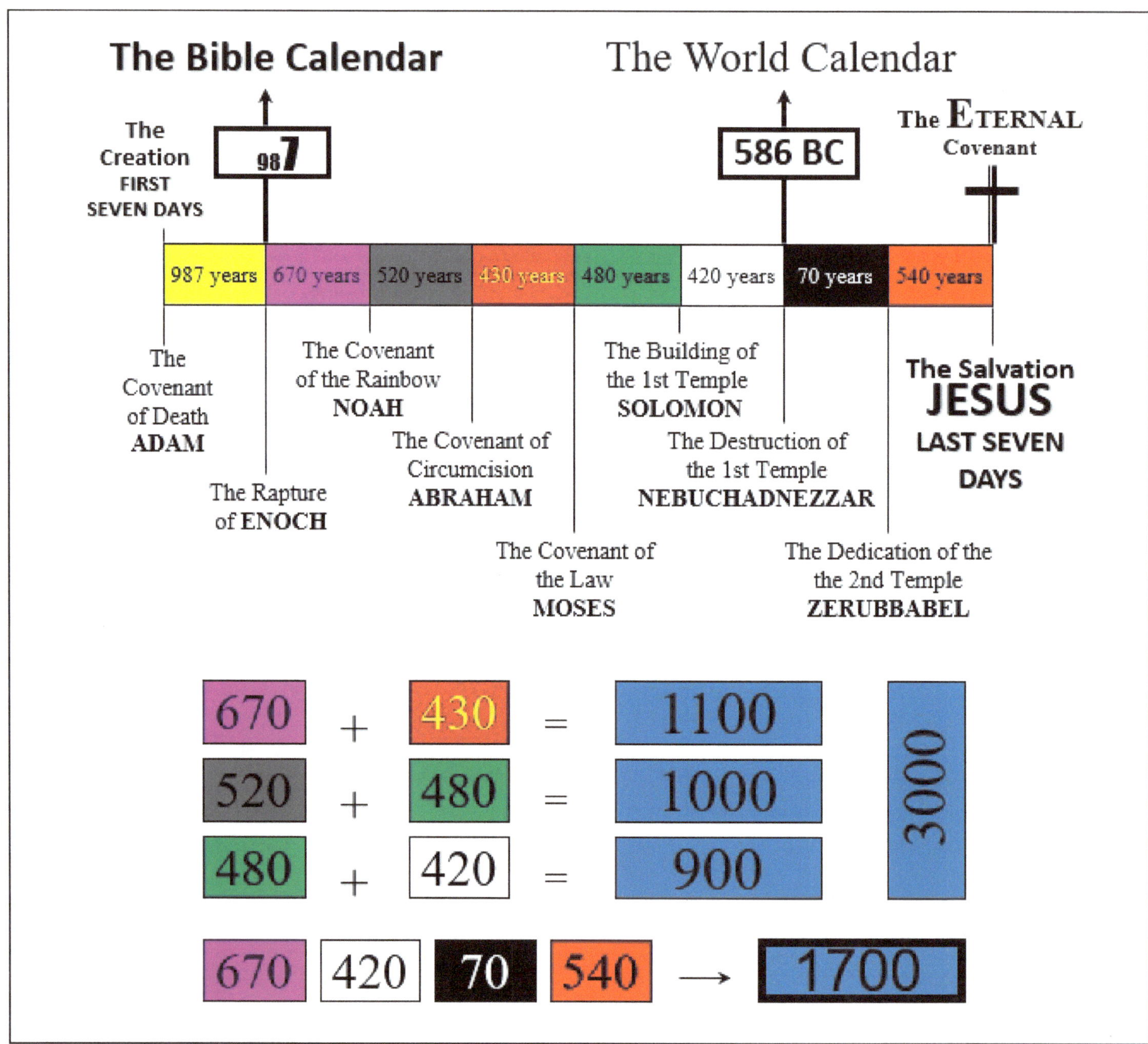

The Bible Chronology Matrix

Jesus says in John 7:6: "My time has not yet come. For you, any time is right" (John 7:6, NIV).

The divine plan for our salvation history, which was the moment of Creation, the moment of Salvation by Adam, Enoch, Noah, Abraham, Moses, and Jesus, had to take **place in a particular year, for a specific month, for a specific week, for a particular day and during a particular hour, minute and second**.

Therefore, I had longed to find out more about the time, times, and durations of what happened when and who lived, when, and even where. We want the truth and nothing but the truth.

Does the world offer truth? No, it teaches us partial truth. Make your thoughts, thinking, views, and perspectives obedient to Christ, just as 2 Corinthians 10:5 says:

"We demolish arguments and every pretension that sets itself up against the knowledge of God, and we take captive every thought to make it obedient to Christ." (2 Corinthians 10:5, NKJV)

During prayer and worship, in a vision, I saw a Bible Chronology that starts with Adam in the Old Testament and reaches through all ages, through the entire Bible. It bridges the Old with the New Testament leading right to Jesus, his birth, his death, and his resurrection. The scripture promises the covenant of Jesus' blood gives eternal life to whoever kneels before his name.

THE MATHEMATICAL SYSTEM OF THE BIBLE CHRONOLOGY deals with the text of Genesis 5:

"When Adam had lived 130 years, he had a son in his likeness, in his image, and he named him Seth. After Seth was born, Adam lived 800 years and had other sons and daughters. Altogether, Adam lived 930 years, and then he died. When Seth had lived 105 years, he became the father of Enosh. After he became the father of Enosh, Seth lived 807 years and had other sons and daughters. Altogether, Seth lived 912 years, and then he died. When Enosh had lived for 90 years, he became the father of Kenan. After he became the father of Kenan, Enosh lived 815 years and had other sons and daughters. Altogether, Enosh lived 905 years, and then he died." (Genesis 5:3-11, NIV)

From studying these lines, I created the BIBLE CHRONOLOGY TABLE. I started this work full-time for one year, from June 1995 to June 1996. I marked down all Bible verses dealing with time and chronologies throughout the entire Bible, from Genesis to Revelation. First, I started with a paper notebook. Then, I copied the lines into the Microsoft Word 5.5 processing software from 1991. Then I stored the file on the IBM 5170 (IBM AT) and 5 1⁄4-inch Floppy Disks. Soon, I copied from 5 1⁄4-inch to 3 1⁄2-inch disks, then to the Iomega Zip- and Jaz-Drive, the to different external harddisks. The revised version of this work is stored in the.

During a further prayer and worship session, I received a vision of **a Mathematical Numerical relationship between the number 430 mentioned in Galatians 3:17 and the number 670 from Genesis 5. Both of these multiples of ten numbers add up to a multiple of a hundred number. 430 stands for Moses, saved in a basket – Hebrew tebat (תֵּבַת). 670 stands for Noah, protected in a boat – Hebrew tebat (תֵּבָה).**

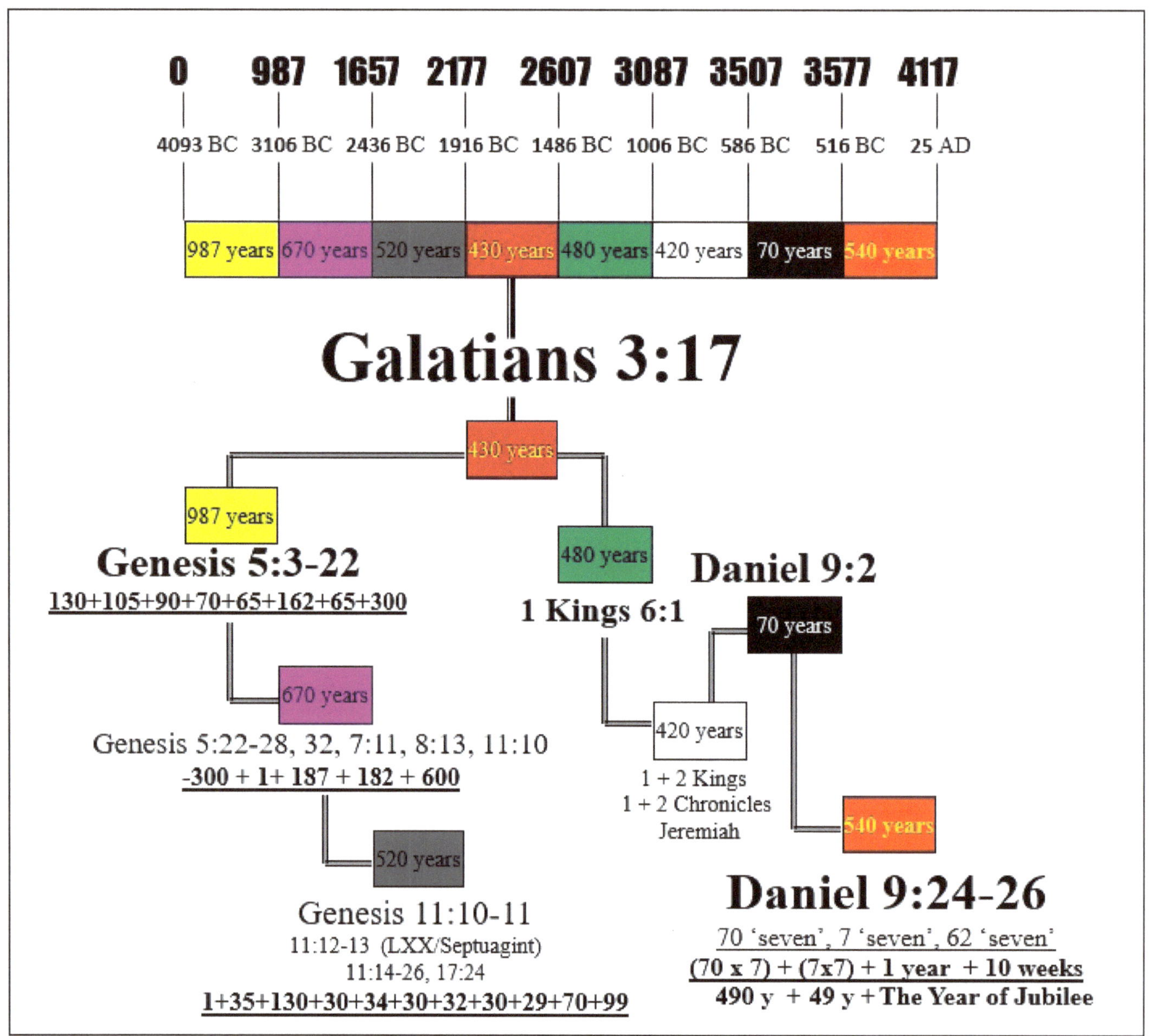

"I mean this: The law, introduced 430 years later, does not set aside the covenant previously established by God and thus do away with the promise." (Galatians 3:17, NIV)

The vision showed me that the Bible Chronology comprises numbers that reach from covenant to covenant. I found that the Bible Chronology is made of multiples of ten (10, 20, 30,…) except the first two numbers 0 and 987.

0-987-670-520-430-480-420-70-540

A string of numbers **0—987—670—520—430—480—420—70—540** defines all ages from Adam to Jesus from covenant to covenant. Altogether there are **4117 years** from Creation to Salvation.

The Mathematical System of the Bible Chronology makes this statement: Partially direct accessibly and partially hidden are eight numbers to be found in the Bible: 987, 670, 520, 430, 480, 420, 70, 540. These "Chronology numbers" match into a Mathematical System like a "riddle," which defines the Bible Chronology from Creation to Salvation. **These sets of numbers are structured to form pairs made of multiples of ten that sum up into multiples of a hundred–except the first number 987.** When allocated in the correct order, these pairs define the Bible Chronology from Adam's Creation to Salvation by Jesus.

- 0 years at the Creation (Covenant of Life and Death with Adam)
- 987 years are from Adam to Enoch (The Rapture of Enoch)
- 670 years are from Enoch to Noah (Covenant of Rainbow)
- 520 years are from Noah to Abraham (Covenant of Circumcision)
- 430 years are from Abraham to Moses (Covenant of Law)
- 480 years are from Moses to Solomon (Covenant of the Temple)
- 420 years are from Solomon to Nebuchadnezzar (End of Covenant of the Temple)
- 70 years are from Nebuchadnezzar to Zerubbabel (Covenant of the Temple II)
- 540 years are from Zerubbabel to Jesus (Covenant of the Blood of Jesus)

The Bible Chronology puts in a grid that erases all accrued gaps and inaccuracies and provides the exact number of years when it was time for Jesus to go to Jerusalem.

"No one who wants to become a public figure acts in secret. Since you are doing these things, show yourself to the world." For even his brothers did not believe in him. Therefore Jesus told them, "My time is not yet here; for you, any time will do. The world cannot hate you, but it hates me because I testify that its works are evil. You go to the festival. I am not going up to this festival because my time has not yet fully come." After he had said this, he stayed in Galilee. (John 7: 4-9, NKJV)

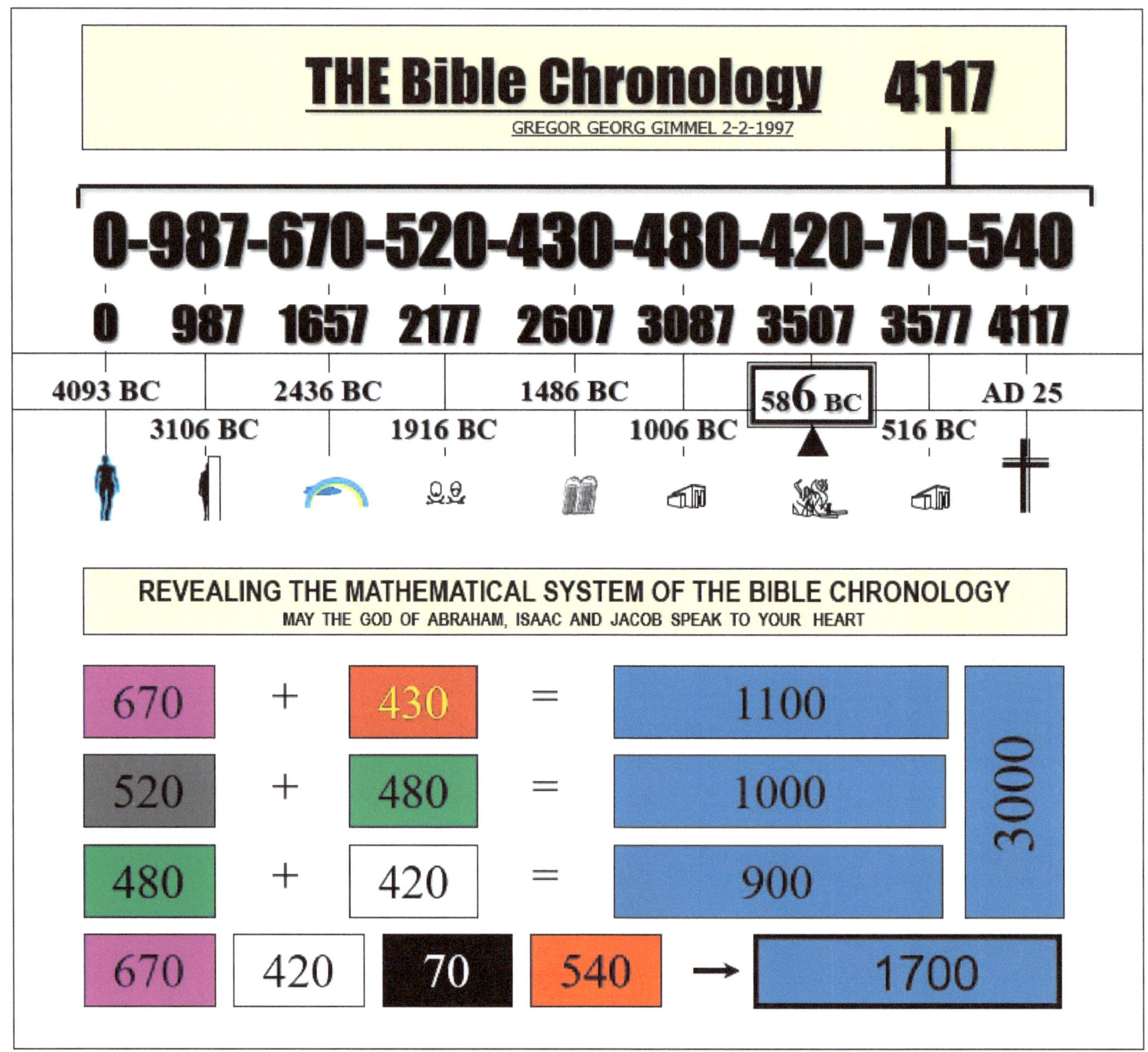

THE MATHEMATICAL SYSTEM OF THE BIBLE CHRONOLOGY shows that God has a plan since the beginning of time. He is always on time. The whole scripture is a big puzzle set. With dedication, prayer, and worship, God shows all mysteries and miracles unknown at first sight. These puzzle pieces matched together with an accurate structure by a mathematical system when the Creation was, when Noah's Flood happened, in what year BC/AD Adam, Noah, Abraham, Moses, Solomon, David, and Jesus lived on this planet.

3. THE BIBLE CHRONOLOGY TABLE

This is the original BIBLE CHRONOLOGY TABLE, which I started in 1995 and completed in 1997. I revised this calendar in 2012 after realizing that the crucifixion year of Jesus is AD 25 and not AD 24. I modified it again in 2021 after describing the five ways to calculate the earthly life of Jesus.

In the original 1995 calendar, I did not factor in no year zero in the historical anno domini calendar. The year 1 BC is followed by AD 1 in the historical calendar.

On the other hand, there is a year zero in the astronomical calendar. The astronomical year zero is the same as the historical year 1 BC.

Hence, the year of the conjunction of Saturn and Jupiter (Star of Bethlehem) is the astronomical year 7 BC (according to Johannes Kepler), which is the historical year 8 BC.

This is in line with scripture saying Herod killed the babies two years and under.

Jesus was 2 years old when the Magi came from the East and visited him in late summer / the fall of 8 BC. Jesus was born in late summer or early fall of 10 BC.

The Bible Chronology Table is based on the 1984 Edition of the New International Version (NIV). Since around 2010, I have started to prefer the NKJV over the NIV. I understand the original 1978 and 1984 NIV editions were more accurate than later editions.

What is the advantage of the NIV over other English Bible translations? The NIV contains the Word of God found in the 280 BC translated Septuagint Bible, which is missing, which are gaps not found in the (more reliable) Masoretic Hebrew Bible.

The Septuagint contains complementary information essential for the BIBLE CHRONOLOGY TABLE.

BIBLE HISTORY AND WORLD HISTORY

GREGOR GEORG GIMMEL

First edition 2-2-1997, 1st revision 6-13-2012, 2nd revision 1-19-2021

reference:

THE HOLY BIBLE, New International Version, Thompson Chain Reference Edition,

Hodder and Stoughton, London/Sydney/Auckland/Toronto,

5th impression, December 1989

brackets [ ] indicate **inaccurate data**

parenthesis ( ) indicate **calculated/revealed data**

~ indicates **estimated data**

y stands for **year**

m stands for **month**

d stands for **day**

This chapter is book-in-book. It contains a copy of the original BIBLE CHRONOLOGY TABLE, which I started in 1995 and completed in 1997. I revised it in 2012, moving the crucifixion year of Jesus from 24AD to 25AD when I realized the difference between the Astronomical Calendar, which does - and the World History Calendar does not contain the year zero. I revised it again in 2021 to explain the five ways to calculate the Birth and Death Year of Jesus.

Font = Arial, Size 10

THE OLD TESTAMENT: (Genesis-Malachi)

I The Creation, II The Time before the Fall of Adam, III The Fall of Adam/The Covenant of Death, IV The Covenant of the Rapture/ IV The Time before the Flood

The Calendar from the Covenant of Life & Death with Adam to the Covenant of the Rapture with Enoch & The Calendar from the Covenant of the Rapture with Enoch to the Covenant of the Rainbow with Noah

These are two overlapping calendars. **The first calendar** was started by Adam and inherited by Enoch. Noah began a second calendar. **There is a one-year gap between those two calendars**. The year-gap is taken into account in the life of Lamech, the father of Noah

Enoch was born in the year 687 after Adam's Creation/Fall year. Adam's Fall occurred shortly after the Creation. The book of Jasher in chapter 3[1] goes into detail about what happened to humanity before the Flood. According to that account, humanity was following the Pre-Flood prophet Enoch (Genesis 5:24, Hebrews 11:5, Jude 1:14). Enoch walked faithfully with God. Many kings on earth followed the teachings of Enoch. Then, Enoch was no more because God took him away. The Pre-Flood Rapture of Enoch took place in the year 9-8-7. (Genesis 5:24, NIV).

By faith, Enoch was taken from this life so that he did not experience death: "He could not be found because God had taken him away." For before he was taken, he was commended as one who pleased God. (Hebrew 11:5, NIV) Enoch, the seventh from Adam, prophesied about them: "See, the Lord is coming with thousands upon thousands of his holy ones (Jude 1:14, NIV).

The Rapture of Enoch in 987 symbolizes an end of an era. Death entered mankind through Adam. Escape from death was provided to the prophet Enoch. Enoch's rapture stands for the Rapture of the Church, which will occur before the "Flood". The "Flood" stands for the "Seven Year Tribulation" time. In the Bible Chronology, the last cipher 7 of 9-8-7 is preserved with each consecutive covenant. After 987, every consecutive covenant follows mathematically as a multiple of ten number. 987: 670, 520, 430, 480, 420, 70, and 540 years. These seven 'multiple of ten' numbers stand for seven eras from the covenant of life and death with Adam through the covenant of eternal life with Jesus.

#	Covenant Name	Representative	Years	Bible Calendar	World
1	Covenant of Life and Death	Adam	**0**	0	4093 BC
2	Covenant of Rapture	Enoch – Ge 5	**987**	987	3106 BC
3	Covenant of the Rainbow	Noah – Ge 6-9	**670**	1657	2436 BC
4	Covenant of Circumcision	Abraham – Ge 10-24	**520**	2177	1916 BC
5	Covenant of the Law	Moses – Gal 3:17	**430**	2607	1486 BC
6	Covenant of the Temple I	Solomon – 1 Ki 6:1	**480**	3087	1006 BC
7	Covenant of the End of Temple I	Nebuchadnezzar	420	3507	**586 BC**

[1] The Bible refers to the Book of the Upright (Jasher) in Joshua 10:13 and 2 Samuel 1:18. The Book of Jasher at Sacred-Texts.com, printed first in Hebrew and published in 1613 in Venice, Italy by Yosèf ben Samuel. The Hebrew text was bought, translated and published by the Mormon (Non-Trinity-Non-Christian) J. H. Parry & Company and published in English in 1887. Sepir Ha Yasher, the Hebrew title of this book, means the 'Book of the Upright'. https://www.sacred-texts.com/chr/apo/jasher/3.htm

8	Covenant of the Temple II	Zerubbabel – Da 9:2	**70**	3577	516 BC
9	Covenant of Eternal Life	Jesus – Da 9:24-26	**540**	4117	25 AD

The 420 years from Solomon to Nebuchadnezzar are not provided in a single Bible verse. It was only by prayer, by God's Power to bring this number and this work to life.

I The Creation, II The Time before the Fall of Adam, III The Fall of Adam/The Covenant of Death, IV The Covenant of the Rapture/ IV The Time before the Flood

The son of God, creator of all	Ge 1:1-2:25, Lk 3:38, Ac 3:15		**ETERNITY**
The creation week thrgh JESUS	**Ge 1:26, 2:7, Ex 20:11, 31:17**	***0y*** 0	**4093 BC**
The creation of the universe and the earth	Ge 1:1	0	4093 BC
The creation of a water expanse (canopy)	Ge 1:6	0	4093 BC
The creation of a one-land/one-sea earth	Ge 1:9	0	4093 BC
The earth watered from underground	Ge 2:5	0	4093 BC
No rain (on the pre-flood earth)	Ge 2:5	0	4093 BC
The creation of all animals including	Ge 1:21, Job 3:8, 7:12, 41:1-10,	0	4093 BC
(dinosaur-) sea monsters/ (=Leviathan)	Ps 74:13-14, 104:25-26, 148:7,	0	4093 BC
The creation of all animals including	Isa 27:1, Eze 32:2, Rev 13:1	0	4093 BC
(dinosaur-) land monsters (=Behemoth)	Ge 1:24, 7:8, Job 40:15-24,	0	4093 BC
The creation of Adam & Eve	Ge 1:26, 2:7, 2:18, 21-24, 1 Ti 2:13	0	4093 BC
The garden in the East, in Eden	Ge 2:8	0	4093 BC
Death Makes Time a Value	Ge 3:22, Ro 5:14	0	4093 BC
The Cov. Life/Death, ADAM, 0y	**Ge 2:17, 3:6, 3:23, Hos 6:7, Ro 5:12**	**0**	**4093 BC**
The life of Adam 930y	Ge 5:3-5, Lk 3:38	0- 930	4093-3163 BC
The birth of Seth, Adam 130y	Ge 5:3-5, Lk 3:38	130y 130	3963 BC
The life of Seth 912y	Ge 5:3-8, Lk 3:38	130-1042	3963-3051 BC
The birth of Enosh, Seth 105y	Ge 5:3-8, Lk 3:38	105y 235	3858 BC
The life of Enosh 905y	Ge 5:6-11, Lk3:38	235-1140	3858-2953 BC
The birth of Kenan, Enosh 90y	Ge 5:9-14, Lk 3:37	90y 325	3768 BC
The life of Kenan 910y	Ge 5:9-14, Lk 3:37	325-1235	3768-2858 BC
The birth of Mahalalel, Kenan 70y	Ge 5:12-17, Lk 3:37	70y 395	3968 BC
The life of Mahalalel 895y	Ge 5:12-17, Lk 3:37	395-1290	3968-2803 BC
The birth of Jared, Mahalalel 65y	Ge 5:15-20, Lk 3:37	65y 460	3633 BC
The life of Jared 962y	Ge 5:15-20, Lk 3:37	460-1422	3633-2671 BC
The birth of Enoch, Jared 162y	Ge 5:18-24, Lk 3:37	162y 622	3471 BC
The life of Enoch 365y	Ge 5:18-24, Lk 3:37	622-987	3471-3106 BC
The birth of Methuselah, Enoch 65y	Ge 5:21-27, Lk 3:37	65y 687	3406 BC
The life of Enoch aft. Methus.'s birth 300y	Ge 5:22, Lk 3:37	300y 687-987	3406-3106 BC
The Rapture of Enoch, EN 365y	**Ge 5:24, Jude:14, Heb 11:5 = 987y**	**987**	**3106 BC**
I The Time after the covenant of the Rapture/ The Time of Noah/ The Covenant of the Rainbow			
The Rapture of Enoch, EN 365y	**Ge 5:24, Jude:14, Heb 11:5 = 987y**	**987**	**3106 BC**
The life of Enoch* aft. Methus.'s birth 300y	Ge 5:22, Lk 3:37	-300y 687-987	3406-3106 BC
The life of Methuselah 969y	Ge 5:21-27, Lk 3:37	687-1656	3406-2437 BC

The birth of Lamech, Methuselah 187y	Ge 5:21-27, Lk 3:37	187y 874	3219 BC
---The one-year gap* of the calendar	Ge 5:25-32, 7:11	1y*874-875	3219-3218 BC
The life of Lamech 777y	Ge 5:25-31, Lk 3:36	875-1652	3218-2441 BC
The birth of Noah*, Lamech 182y	Ge 5:28-32,6:6,9:28-29, Lk 3:36	182y 1057	3036 BC
The life of Noah 950y	Ge 5:28-32,6:6,9:28-29, Lk 3:36	1057-2007	3036-2086 BC
The birth of Shem, Noah 500y+1y*	Ge 5:27, 8:13, 11:10, 5:32	1558	2535 BC
The life of Shem 600y	Ge 7:6, 11:10, [5:22], Lk 3:36	1558-2158	2535-1935 BC
The Destruction of the Earth:	Ge 6:13	1656	2437 BC
of the expanse (water canopy destr)	Ge 6:13, 7:11	1656	2437 BC
of the one-land earth (continental drift)	Ge 6:13, 7:11	1656	2437 BC
of all animals (fossilized, petrified)	Ge 6:7, 7:21-23	1656	2437 BC
of all mankind (fossilized, petrified)	Ge 6:13, 7:21-23	1656	2437 BC
Man's life to be cut short to 120y	Ge 6:3, Job 14:5,	1656	2437 BC
The prophecy of mankind's life on earth will be limited to 120 years (Genesis 6:3)			**2437 BC**
The Transformation of the Earth is completed with the death of Moses (Ex 7:7, Dt 31:2, 34:7)			
Hence, the Ice Age ended with the end of Moses life on earth			
The Ice Age: (Noah-~Moses) ~950 years	**Ge 6:3, 7:11 Ex 7:7, Dt 31:2, 34:7**	**1656-~2606**	**2436-~1486BC**
The Causing of polar cooling/ Snow Fall/	Ge 7:11, Job 6:16, 24:19, 37:6, 38:22	1656-	2437- BC
Man's life to be prolonged again (1000y K)	Isa 65:20, Rev 20:2-7		
Noah enters the arch, Noah 599y1m9d	Ge 7:10,11, Mt 24:38, Lk 17:26	1656	2437 BC
The flood arrives, Noah 599y1m16d	Ge 7:11, Job 22:16, Mt 24:39, Lk 17:27	1656	2437 BC
The death of mankind and Methuselah	Ge 5:27, 6:7, 1 Pe 3:18-21, 2 Pe 2:5	1656	2437 BC
The flood dried up, Noah 600y	Ge 8:13	600y* 1657	2436 BC
Noah leaves the arch, Noah 600y1m26d	Ge 8:14-17	(1m26d) 1657	2436 BC
The Coven.of Rainbow, N 600y+	**Ge 9:8-17, 8:13, 7:6, 9:28**	**= 670y 1657**	**2436 BC**

The Calendar from the Covenant of Rainbow to the Covenant of Circumcision

This calendar was started by Noah and inherited by Shem. Genesis 1-11 was copied by Abraham, living with Shem for 39 years in the House of Noah in the Mountains of Ararat as by the Book of the Upright/Jasher. Cainan is the son of Arphaxad and the father of Shelah. The Masoretic Text is missing this generation. The lost generation of Cainan, son of Arphaxad, is captured in Luke's Gospel, chapter 3. In this Calendar, **Cainan of Luke 3:36** is **130 years to Shelah** in the NIV New International Version **to complete the 520 years from Rainbow to Circumcision**. Only Genesis 11:13b LXX is used to fill this gap.

Genesis 11:13b (NIV/LXX) "When Cainan had lived **130 years**, he became the father of Shelah. And after he became the father of Shelah, Cainan lived **330 years** and had other sons and daughters." (Genesis 11:13b Greek Septuagint LXX, NIV, New International Version).

Και (And) έζησε (lived) Καϊνάν (Cainan) εκατόν (a hundred) και (and) τριάκοντα (thirty) έτη (years) και (and) εγέννησε (he procreated) τον (him) Σαλά (Salah) και (and) έζησε (lived) Καϊναν (Cainan) μετά (after) το (the) γεννήσαι (procreating) αυτω (this) τον (him) Σαλα (Salah) έτη (years) τριακόσια (three hundred) τριάκοντα (thirty) και (and) εγέννησεν (he procreated) υιούς (sons) και (and) θυγατέρας (daughters) και (and) απέθανε (he died).

I The Covenant of the Rainbow/ The Time after the Flood

The Covenant of the Rainbow, N, 600y+	**Ge 9:8-17, 8:13, 7:6, 9:28**	**=670y 1657**	**2436 BC**
The birth of Arphaxad, Shem 100y	Ge 11:10, 8:13-14	1y 1658	2435 BC
The life of Arphaxad 465y	Ge 11:10-13 (LXX), Lk 3:36	1658-2123	2435-1970 BC
The Prophecy of Canaan	**Ge 9:25 → 1 Ki 9:20-23**	**~1693**	**~2400 BC**
The birth of Cainan, Arphaxad 35y	Ge 11:12-13 (LXX), Lk 3:36	35y 1693	2400 BC
The life of Cainan 460y	Ge 11:12-13 (LXX), Lk 3:36	1693-2153	2400-1940 BC
The birth of Shelah, Cainan 130y	Ge 11:12-15 (LXX), Lk 3:35	130y 1823	2270 BC
The life of Shelah 433y	Ge 11:12-15 (LXX), Lk 3:35	1823-2256	2270-1837 BC
The birth of Eber, Shelah 30y	Ge 11:14-17, Lk 3:35	30y 1853	2240 BC
The life of Eber 464y	Ge 11:14-17, Lk 3:35	1853-2317	2240-1776 BC
Tower of Babel (Peleg)	Ge 10:25, 11:8,16	1887	2206BC
Peleg:(earth divided=different language spoken by the dispersed nations)		1887	2206BC
The destruction of the one-language world	Ge 11:1, 7, 9	1887	2206 BC
The dispersion of the nations	Ge 11:8, 9	1887	2206 BC
The (genetic) separation of the nations/	Ge 11:8, 9, Nu 12:1, 1 Ki 10:1-13	1887	2206 BC
different sun radiation/vegetation/nutrition	Ge 11:8, 9, Mt 12:42, Lk 11:31,	1887	2206 BC
causing national features (races)	Ge 11:8, 9, Ac 8:27	1887	2206 BC
Hamitic: Hittites: (Cathay)	Ge 11:8, 9	1887	2206 BC
-Asian, Eskimo, Nat.Amer.Ind/Australian)	Ge 11:8, 9	1887	2206 BC
-Ancient Europe: (Etruscans, Basque)	Ge 11:8, 9	1887	2206 BC
-Ancient Mesopotamia: (Sumerians)	Ge 11:8, 9	1887	2206 BC
Hamitic: Sinites:(China)	Ge 11:8, 9	1887	2206 BC
Hamitic: Put, Cush: (Africans)	Ge 11: 8, 9, Nu 12:1, Ac 8:27	1887	2206 BC
..Shemitic: Eber: (Hebrew, Israel),Arabians	Ge 11:8, 10:24, 11:15-16	1887	2206 BC
..Japhetic: (Indo-European):	Ge 11:8	1887	2206 BC
-India,Europe,Americas,S.Africa,Australia)	Ge 11:8	1887	2206 BC
The birth of Peleg, Eber 34y	Ge 11:16-19, Lk 3:35	34y 1887	2206 BC
The life of Peleg 239y	Ge 11:16-19, Lk 3:35	1887-2126	2206-1967 BC
The birth of Reu, Peleg 30y	Ge 11:18-21, Lk 3:35	30y 1917	2176 BC
The life of Reu 237y	Ge 11:18-21, Lk 3:35	1917-2154	2176 BC
The birth of Serug, Reu 32y	Ge 11:20-23, Lk 3:35	32y 1949	2144 BC
The life of Serug 230y	Ge 11:20-23, Lk 3:35	1949-2179	2144-1914 BC
The birth of Nahor, Serug 30y	Ge 11:22-25, Lk 3:34	30y 1979	2114 BC
The life of Nahor 148y	Ge 11:22-25, Lk 3:34	1979-2127	2114-1966 BC
The birth of Terah, Nahor 29y	Ge 11:24-26, 32, Lk 3:34	29y 2008	2085 BC
The life of Terah 205y	Ge 11:24-26, 32, Lk 3:34	2008-2213	2085-1880 BC
The birth of Abram, Terah 70y	Ge 11:26, 12:4, 25:7, Lk 3:34	70y 2078	2015 BC
The life of Abram/Abraham 175y	Ge 11:26, 12:4, 25:7, Lk 3:34 2015-1840 BC		2078-2253
The Cov. of Circumc., ABR. 99y	**Ge 17:1, 24, Lev 12:3**	**+99y = 520y 2177**	**1916 BC**

The birth of Isaac, Abraham 100y	Ge 17:21, 21:5, 35:28	2178	1915 BC
The life of Isaac 180y	Ge 17:21, 21:5, 35:28	2178-2358	1915-1735 BC

The Calendar from the Covenant of Circumcision to the Covenant of the Law In this Calendar, Galatians 3:17 determines the multiple-of-ten 430 years from Circumcision to the Law on Mount Sinai. The four *extra-long* generations *Levi, Kohath, Izhar, Korah,* determine Israel's stay in Egypt to 239 years, which are *ten short generations (10 x 23.9 years).*

I The Time of the Patriarchs, II The Time of the Building of the Israel Nation

The Cov. of Circumc, ABR. 99y	**Ge 17:1, 24, Lev 12:3 +99y = 520y**	**2177**	**1916 BC**
The birth of Isaac, Abr.100y, 1y af.cc.cov	Ge 17:1, 21, 21:5, 35:28, Lk 3:34 *1y*	2178	1915 BC
The life of Isaac 180y	Ge 17:21, 21:5, 35:28, Lk 3:34	2178-2358	1915-1735 BC
The birth of Jacob/Israel, Isaac 60y	Ge 25:26, 47:28, Lk 3:34 *60y*	2238	1855 BC
The life of Jacob/Israel 147y	Ge 25:26, 47:28, Lk 3:34	2238-2385	1855-1708 BC
The birth of Joseph, Jacob 91y	Ge 29:31-30:24	2329	1764 BC
The life of Joseph 110y	Ge 41:29, 46-47, 53-54, 45:11, Ge 47:9, 50:22	2329-2439	1764-1654 BC
Joseph invents the Hebrew script from Egyptian signs		***2329-2439***	***1764-1654 BC***

→→→As per the research work done by Doug Petrovich/ Sarah K. Doherty, Book "The World's Oldest Alphabet" published by Carta in 2016, Joseph developed out of the Egyptian signs a new Hebrew alphabeth. Joseph and his sons appear to have copied the Book of Genesis from the **Pre-Flood script invented by Adam** into the **Hebrew script invented by Joseph**. It appears that Genesis was first written in Adam's Pre-Flood script. It appears Adam wrote Genesis 1-4 and the first half of chapter 5. Genesis 1-5 was passed on from generation to generation. Adam's Pre-Flood script was learned by Noah and his son Shem. Noah wrote Genesis 6-9. Shem wrote Genesis 10-11. Abraham wrote Genesis 12-24. Isaac wrote Genesis 25-26. Jacob wrote Genesis 27-36. Joseph wrote Genesis 37-50. Abraham grew up in Noah's house according to the Book of the Upright (Jasher) "And Abram was in Noah's house thirty-nine years" (Jasher 9:6, Sacred-Text.Com). Job was a friend of Abraham who moved during the time of Abraham into Syria near Israel on the opposite side of the Lake of Gennesaret to "Hammam Ayyub" (Fountain of Job) in El Shykh Saad, Syria at 32°50'06.4"N 36°02'06.8"E (Mosque of the Rock of Job). Abraham trained his son Isaac in the reading and writing skills of the Pre-Flood script. The skill was taught from generation to generation until Joseph. It appears that God used Joseph to develop a new Hebrew alphabeth made up from Egyptian signs. It appears that with the development of the Hebrew alphabeth, Joseph copied and trans-scripted the entire Book of Genesis (Bere'shit ("in the beginning") = Genesis) from the **Pre-Flood script invented by Adam** into the new **Hebrew script invented by Joseph**. Moses assigned the Levites to copy the Book of Genesis into the Hebrew script. Moses and the Levites wrote the **Four Books of Moses**. (Shemot ("names") = Exodus, Vayiqra ("and he called") = Leviticus, Bemidbar ("in the wilderness") = Numbers and Devarim ("words") = Deuteronomy. Together, the First Five Books of the Bible are called the "Torah" meaning Instruction/Teaching and Law.

Joseph before Pharao, 30y	Ge 41:46	2359	1734 BC
Joseph speaks Egyptian	Ge 42:23	2367	1726 BC
Israel moves to Egypt, Israel 130y	Ge 47:9, 47:1-9, 46:26-27 *130y*	2368	1725 BC
The prophesied 4x100 years in Egypt	Ge 15:13, Ac 7:6	2368-2607	1725-1486 BC
(revealed as 4 generations in Egypt)	Ge 15:16	2368-2607	1725-1486 BC
Levi, Kohath, Izhar, Korah	Nu 16:1, Ge 46:11, Ex 6:18	2368-2607	1725-1486 BC
(100y => 1 generation)	Ge 15:13, 16	2368-2607	1725-1486 BC
The multiplication of Israel	Ac 7:14, Ex 1:5 (Dead Sea Scr)		

from 75 to 603,550 people	Ge 46:27 (LXX), Ex 38:26, Nu 1:46	2368-2607	1725-1486 BC
Israel's Stay in Egypt 239y	***Gal 3:17, Ge47:9, Ge46:8,11 239y***	***2368-2607***	***1725-1486 BC***
	Ex 6:16-20, 1 Ch 6:1-3,		
The birth of Levi, Jacob ~82y	Ge 29:31-30:24	~2320	~1773 BC
The life of Levi 137y	Ge 29:31-30:24, Ex 6:16,	~2320-2457	~1773-1636 BC
The birth of Kohath, Levi ~40y	Ge 46:8, 11, Ex6:16, 1 Ch6:1	~2360	~1733 BC
The life of Kohath 133y	Ex 6:18	~2360-2493	~1733-1600 BC
Kohath moves with Israel to Egypt	***Ge 46:8, 11***	***2368***	***1725 BC***

Kohath is the **grand-father of Moses**. He moves as a child from Canaan to Egypt in 1725 BC. Thus, there are 239 years (and NOT 430 years!) from Israel's move to Egypt until the Exodus of Israel from Egypt.

Israel multiplies during 239 years (NOT 430 years!) (1725-1486 BC) **in 10 generations** with an average generation age of **23.9 years** from 75 or 70 to **603,550 people**. Very similar figures occurred for mankind after the Flood. Mankind multiplied during 230 years (2436-2206 BC) **in 10 generations** with an average generation age of 23.0 years from 8 to over 600,000 people (see Book of Jasher 9:23). Kohath is the grand-father of Moses. He moves as a child from Canaan to Egypt in 1725 BC. Assumption: Koath moves to Egypt as about an 8-year old boy. Kohath's son Amram is born when Kohath is about 80 years old. Amram's son Moses is born when Amram is about 87 years old. Moses is 80 years old when he meets God on Mount Sinai in 1486 BC. These are **the four extraordinary generations** of Egypt from Levi to Moses: *Levi-Kohath-Amram-Moses*. NOTE: If Israel was staying in Egypt 430 years, then the average generation age between Kohath and Moses was 3x143.3 years which is beyond the age of people during this period of time.

The birth of Amram, Kohath ~80y	Ex 6:20, 1 Ch 6:2	~2440	~1653 BC
The life of Amram 137y	Ex 6:20	~2440-2577	~1653-1516 BC
The birth of Moses, Amram ~87y	Ex 7:7, Dt 31:2, 34:7	2527	1566 BC
The life of Moses 120y	Ex 7:7, Dt 31:2, 34:7	2527-2647	1566-1446 BC
The birth of Aaron	Ex 7:7, Nu 33:39	2524	1569 BC
The life of Aaron 123y	Ex 7:7, Nu 33:39	2524-2647	1569-1446 BC
Israel living in Canaan & Egypt 430y	Gal 3:17, Ex 12:40-41 (LXX)	2177-2607	1916-1486 BC
	Ge 15:13-16		
The Exodus, month of Abib, Moses 80y	Ex 12:41, Ex 34:18, Nu 33:3	2607	1486 BC
The law introduced, Mt Sinai	Gal 3:17, Ex 34:1-28, 32:15,16	2607	1486 BC
+++Covenant of the Law, MOSES, 80y	**Gal 3:17 = 430y**	**2607**	**1486 BC**
	Ex 12:40-41 (LXX)		
The Manna & Quail 1m14d	Ex 16:1	2607	1486 BC
At Mount Sinai 3m	Ex 19:1,11,24:18,34:28	2607	1486 BC
The Census 1y1m	Nu 1	2608	1485 BC
Leaving Mount Sinai 1y1m19d	Nu 10:11	2608	1485 BC
12 men sent to explore Canaan for 40 days	Nu 10:11, Jos 14:7,10	2608	1485 BC
Joshua/Hoshea son of Nun 40y old	Jos 14:7-10, Nu 10:11,13:8,25	2608	1485 BC
Israel in the desert 40y, 1 year for each day	Nu 14:33-34,32:13, Dt1:3,2:7,29:5	2607-2647	1486-1446 BC
	Jos5:6, Ac7:36		
The Zered Valley crossed 38y	Dt 2:14	2645	1448 BC

The death of Aaron on Mount Hor 39y4m	Nu 33:38-39, Dt 32:50	2646	1447 BC
The command to leave Horeb 39y10m	Dt 1:3	2646	1447 BC
The king of Heshbon subdued	Dt 2:24	2646	1447 BC
The king of Bashan subdued	Dt 3:1	2646	1447 BC

The death of Moses on Mount Nebo	***Dt 31:2,32:49,34:1,7***	***2647***	***1446 BC***

With the death of Moses at 120years, the prophecy of Genesis 6:3 (120y) gets fulfilled. The Ge 6:3 prophecy is about the transformation of planet earth and its life from Pre- to Post-Flood. The death of Moses marks the completion of this transformation. It falls together with the time of the end of the Ice Age. The Ice Age lasted ~990 years or ~1000 years (2436-1446 BC). The Death of Moses stands for the End of the Ice Age in 1446 BC: During this time, around 1446 BC, Gomer (The Celts), the Oldest son of Japheth, move from the Black Sea into present-day Eastern & Western Europe (Italy, Germany, France, Spain, Portugal, England, Wales, Ireland). Gomer, an Indo-European speaking nation or nations conquer Eastern & Western Europe from Non-Indo-European speaking nations, descendants of Nimrod (Neanderthal, Cro-Magnon, Basque, Pile Dwellers, Picts, Rhaetians, others)

Israel grieves for Moses 30d	Dt 34:8		2647	1446 BC
The crossing of the Jordan	Jos 3:16		2647	1446 BC
The covenant renewed, Mt Ebal	Jos 8:30-35	*40y*	2647	1446 BC

The Calendar from the Covenant of the Law to the Temple Inauguration of Solomon

In this calendar, **1 Kings 6:1** determines the multiple-of-ten 480 years from Law to the Temple. Judge 11:26 – 300 years in Heshbon and Aroer indicates that the Israelites did keep an exact calendar as an integral part of the Torah, the Laws the Holy Scriptures.

I The Time of the United Tribes of Israel, II The Kingdom of Israel

+++Covenant of the Law, MOS, 80y	***Gal 3:17, Ex 12:40-41 (LXX)***		***430y***	***2607***	***1486 BC***
The Zered Valley crossed 38y	Dt 2:14		2y	2645	1448 BC
Israel in the desert 40y=38y+2y	Ac 7:36, Jos 5:6		40y	2607-2647	1486-1446 BC
The crossing of the Jordan	Jos 3:16			2647	1446 BC
The life of Joshua/Hoshea 110y	Jos 14:7,10,24:29, Jdg2:8, Nu13,14Ex17,24,32			2567-2677	1526-1416 BC
Caleb/Joshua explore Canaan at age 40	**Jos 14:7**,10, Nu1:1,10:11,13:8			2607	1486 BC
Crossing of the Jordan, The fall of Jericho,	Jos 3:16, Jos 5:13-6:27			2647	1446 BC
King of Ai subdued, decept. of the Gibeon.	Jos 8:1-29, Jos 9:1-27			2647	1446 BC
The sun stands still; five Amorite kings killed	Jos 10:13, Jos 10:16-28			2647	1446 BC
Southern cities conquered; northern k def.	Jos 10:29-43, Jos 11:1-23			2647	1446 BC
Joshua judge over Israel (30y)	Jos 1:5, 14:7,10 Jdg 2:8	**30y**	***30y***	2647-2677	1446-1416 BC
A new sinning generation (20y)	Jdg 2:10-11,17	**20y**	***20y***	2677-2697	1416-1396 BC
Cushan subdues Israel 8y	Jdg 3:8			2689-2697	1404-1396 BC
Othniel judge over Israel 40y	Jdg 3:11	**40y**	***40y***	2697-2737	1396-1356 BC
Eglon subdues Israel 18y	Jdg 3:14			2719-2737	1374-1356 BC
Ehud judge over Israel 80y	Jdg 3:30	**80y**	***80y***	2737-2817	1356-1276 BC
Shamgar judge over Israel	Jdg 3:31			2817	1276 BC
Jabin subdues Israel 20y	Jdg 4:3			2797-2817	1296-1276 BC
Deborah judge over Israel 40y	Jdg 5:31	**40y**	***40y***	2817-2857	1276-1236 BC
Midian subdues Israel 7y	Jdg 6:1			2850-2857	1243-1236 BC

Gideon judge over Israel 40y	Jdg 8:28	**40y**	***40y***	2857-2897	1236-1196 BC
Abimelech judge over Israel 3y	Jdg 9:2	**3y**	***3y***	2897-2900	1196-1193 BC
Tola judge over Israel 23y	Jdg 10:2	**23y**	***23y***	2900-2923	1193-1170 BC
Jair judge over Israel 22y	Jdg 10:3	**22y**	***22y***	2923-2945	1170-1148 BC
Philistines+Ammonites oppress Israel 18y	Jdg 10:8			2945-2963	1148-1130 BC
Israel occupies Heshbon, Aroer+, 300y	***Jdg 11:26***	***= 300y***		***2645-2945***	***1448-1148 BC***
Jephthah judge over Israel 6y	Jdg 12:7		***6y***	2945-2951	1148-1142 BC
Ibzan judge over Israel 7y	Jdg 12:9		***7y***	2951-2958	1142-1135 BC
Elon judge over Israel 10y	Jdg 12:11		***10y***	2958-2968	1135-1125 BC
Abdon judge over Israel 8y	Jdg 12:14		***8y***	2968-2976	1125-1117 BC
Philistines oppress Israel 40y	Jdg 13:1			2976-3016	1117-1077 BC
Samson judge over Israel 20y	Jdg 15:20,16:31		***20y***	2976-2996	1117-1097 BC
The life of Eli 98y	1 Sam 4:15			2878-2976	1215-1117 BC
Eli judge over Israel 40y	1 Sam 4:18			2936-2976	1157-1117 BC
The Philistines capture the Ark	1 Sam 4:11			2976	1117 BC
The death of Eli	1 Sam 4:18			2976	1117 BC
The Ark in Kiriath Jearim 20y	1 Sam 7:2			2976-2996	1117-1097 BC
Samuel subdues the Philistines	1 Sam 7:2,11			2996	1097 BC
Samuel judge over Israel	1 Sam 7:15		***7y***	2996-3003	1097-1090 BC
The life of Saul 70y	Ac 13:21, 1Sam13:1			2973-3043	1120-1050 BC
Saul king over Israel 40y [42y]	Ac 13:21,[1Sam13:1]		***40y***	3003-3043	1090-1050 BC
The death of Saul	1 Sam 31:4			3043	1050 BC
The life of Ish-Bosheth 42y	2 Sam 2:10, 4:6			3003-3045	1090-1048 BC
Ish-Bosheth king over Israel 2y	2 Sam 2:10			3043-3045	1050-1048 BC
David king over Judah in Hebron 7y6m	2 Sam 2:11, 1Ki2:11,1Ch29:26		***7y(6m)***	3043-3050	1050-1043 BC
David king over Israel in Jerus. 33y	2 Sam 5:4,5,1Ki2:11,1Ch3:4		***33y***	3050-3083	1043-1010 BC
The life of David 70y	2 Sam 5:4			3013-3083	1080-1010 BC
The Ark brought to Jerusalem	2 Sam 6:1,11,12			3050	1043 BC
The death of David	1 Ki 2:10			3083	1010 BC
Solomon king over Israel 40y	1 Ki 2:12,11:42,2Ch9:30			3083-3123	1010-970 BC
Solomon starts building the Temple	1 Ki 6:1, 480th Exodus [1 Ki 6:1, LXX: 440th]			3086	1007 BC
Solomon 4th king Israel	2 Ch 3:2		***3y***	3086-3093	1007-1000 BC
The Cov. of the Temple, SOL	**1 Ki 6:1**		***+1y* = 480y**	**3087**	**1006 BC**
(Temple under construction 1 year)					

The Calendar from the Covenant of the Temple to the End of the Temple by Solomon
Gaps of times when there was no king in Israel and Judah make this the most challenging calendar
The Time of the Kingdom of Israel, II The Time of the Two Separated Kingdom Israel & Judah

The Covenant of the Temple, Solomon		**1 Ki 6:1**		**480y**	**3087**	**1006 BC**
(Temple under construction 1 year)						
The Temple of the LORD completed 7y		1 Ki 6:38			3093	1000 BC
Solomon building his palace 13y		1 Ki 7:1,9:10			3093-3106	1000-987 BC
The prophecy of Canaan fulfilled!		***Ge 9:25 → 1 Kings 9:20-23***			***~3106***	***~987 BC***
~1413 years later: ~1693 (2400 BC) → 3106 (987 BC) = ~1413 years until fulfillment of the prophecy						
The death of Solomon		1 Ki 11:43	36y	***36y***	3123	970 BC
Jeroboam king Israel 21y		1 Ki 15:25	20y		3123-3144	970-949 BC
Nadab king Israel 2y,	2nd Asa	1 Ki 15:25	1y		3143-3144	950-949 BC
Baasha king Israel 24y,	3rd Asa	1 Ki 15:33	24y		3144-3168	949-925 BC
Elah king Israel 2y,	26th Asa	1 Ki 16:8			3167-3168	926-925 BC
Elah murdered,	27th Asa	1 Ki 16:10			3168	925 BC
Zimri king Israel 7d,	27th Asa	1 Ki 16:15			3168	925 BC
---Israel without king for 4years		1 Ki 16:18-22		**4y**	**3168-3172**	**925-921 BC**
Omri king Israel 12y,	31st Asa	1 Ki 16:23		7y	3172-3184	921-909 BC
Ahab king Israel 22y,	38th Asa	1 Ki 16:29		19y	3179-3201	914-892 BC
Ahaziah king Israel 1y,	17th Jehos.	1 Ki 22:51		1y	3198-3199	895-894 BC
Joram king Israel 12y,	18th Jehos.	2 Ki 3:1, [2 Ki 1:17]		12y	3199-3211	894-882 BC
Jehu king Israel 28y		2 Ki 10:36		28y	3211-3239	882-854 BC
Jehoahaz king Israel 17y,	23rd Joash	2 Ki 13:1		14y	3239-3256	854-837 BC
Jehoash king Israel 16y,	37th Joash	2 Ki 13:10		15y	3253-3269	840-824 BC
Jeroboam king Israel 41y,	15th Amaziah	2 Ki 14:23		41y	3268-3309	825-784 BC
---Israel without king for 21 years		**--- 21y**		**21y**	**3309-3330**	**784-763 BC**
Zechariah king Israel 6m,	38th Uzziah	2 Ki 15:8		1y(6m)	3330-3331	763-762 BC
Shallum king Israel 1m,	39th Uzziah	2 Ki 15:13		(1m)	3331	762 BC
Menahem king Israel 10y,	39th Uzziah	2 Ki 15:17		10y	3331-3341	762-752 BC
Pekahiah king Israel 2y,	50th Uzziah	2 Ki 15:23		1y+2y	3342-3344	751-749 BC
Pekah king Israel 20y,	52nd Uzziah	2 Ki 15:27		20y	3344-3364	749-729 BC
Hoshea king Israel 9y,	20th Jotham	2 Ki 15:30,[2 Ki 17:1]		8y	3364-3372	729-721 BC
The Fall of Samaria		***2 Ki 17:5,2 Ki 18:10***		***249y***	***3372***	***721 BC***
The life of Rehoboam 58y		1 Ki 14:21			3082-3140	1011-953 BC
Rehoboam king Judah 17y		1 Ki 11:43,2 Ch 12:13	17y	***17y***	3123-3140	970-953 BC
The death of Rehoboam		2 Ch 13:20			3140	953 BC
Abijah king Judah 3y,	18th Jerob.	1 Ki 15:1, 2 Ch 13:1	2y	***2y***	3140-3143	953-950 BC
Asa king Judah 41y,	20th Jerob.	1 Ki 15:9,2	40y	***40y***	3142-3183	951-910 BC
		Ch 14:1,16:13				
Jehoshaphat k Judah 25y,	4th Ahab	1 Ki 22:41,2 Ch 20:31	21y	***21y***	3182-3207	911-886 BC
Jehoram king Judah 8y,	5th Joram	2 Ki 8:16,2 Ch 21:5, 20	7y	***7y***	3203-3211	890-882 BC

Ahaziah king Judah 1y,	12th Joram	2 Ki 8:25[9:29],2Ch22:2	1y	***1y***	3210-3211	883-882 BC
Ataliah queen Judah 6y		2 Ki 11:3, 2 Ch 22:10,12	6y	***6y***	3211-3217	882-876 BC
Joash king Judah 40y,	7th Jehu	2 Ki 12:1,2 Ch 24:1	37y	***37y***	3217-3257	876-836 BC
Amaziah king Judah 29y,	2nd Jehoash	2 Ki 14:1,2 Ch 25:1	29y	***29y***	3254-3283	839-810 BC
---Judah without king for 10 years		**--- 10y**	**10y**	**10y**	**3283-3293**	**810-800 BC**
Isaiah prophet		Isaiah 1:1			3293-3396	800-697 BC
Uzziah/Azar. k Judah 52y,	27th (26th)Jerob.	2 Ki 14:21,15:1,2Ch26:3	52y	***52y***	3293-3345	800-748 BC
Jotham king Judah 16y,	2nd Pekah	2 Ki 15:32,2Ch27:1	15y	***15y***	3345-3361	748-732 BC
Ahaz king Judah 16y,	17th Pekah	2 Ki 16:1,2Ch28:1	7y	***7y***	3360-3367	733-726 BC
Hezekiah king Judah 29y,	3rd (4th) Hoshea	2 Ki 18:10,1,2Ch29:1	29y	***29y***	3367-3396	726-697 BC
Judah remaining kingdom					3374	719 BC
Manasseh king Judah 55y		2 Ki 21:1,2Ch33:1	55y	***55y***	3396-3451	697-642 BC
Amon king Judah 2y		2 Ki 21:19,2Ch33:21	2y	***2y***	3451-3453	642-640 BC
Josiah king Judah 31y		2 Ki 22:1,Jer25:1-3	31y	***31y***	3453-3484	640-609 BC
Jehoahaz king Judah 3m		Jer 25:1-3,2Ch36:2		***(3m)***	3484	609 BC
Jehoiakim/Eliakim king Judah 11y		Jer 25:1-3,2Ch36:5	11y	***11y***	3484-3495	609-598 BC
Jehoiachin king Judah 3m10d		2 Ki 24:8,2Ch36:9 (3m10d)		***(3m10d)***	3495	598 BC
Zedekiah king Judah 10y4m (11y)		Jer 1:3,2Ki24:18,2Ch36:11	11y	***11y***	3495-3506	598-587 BC
The fall of Jerusalem		Jer 52:5-7,2Ki25:2-4	383y		3506	587 BC
The 1st Temple of the LORD burned down		Jer 52:13,2Ki25:8-9			3507	586 BC
The Cov of the Temple ended NEB		***Jer 52:13, 2 Ki 25:8-9 + 1y = 420y***			***3507***	***586 BC***
The sin of the house of Judah 40y		Eze 4:6			3102-3142	991-951 BC
The sin of the house of Israel 390y		Eze 4:5			3117-3507	976-586 BC

I The Two Separated Kingdoms Israel & Judah, II The Exile of Israel & Judah
III The Time after the Exile of Israel and Judah, IV The Covenant of Crucifixion/ The Eternal Covenant

Tiglath-Pileser subdues Naphtali+	2 Ki 15:29	~3350-3364	~743-729 BC
Shalm. besieges Samaria 3y	2 Ki 18:9-10	3370-3372	723-721 BC
The Fall of Samaria	2 Ki 17:5	3372	721 BC
Sennacherib attacks Judah, 14th Hezk.	2 Ki 18:13	3380	713 BC
Josiah king Judah 31y	2 Ki 22:1,Jer25:1-3	3453-3484	640 BC
Jehoahaz king Judah 3m	Jer 25:1-3,2Ch36:2	3484	609 BC
Jehoiakim king Judah 11y	Jer 25:1-3,2Ch36:5	3484-3495	609-598 BC
Neb. besieges Jehoiakim, 3rd Jehk.	Da 1:1	3488	605 BC
Daniel sent to Babylon	Da 1:3-6, Jer25:11	3488	605 BC
Neb. defeats Pharao, 4th Jehk.	Jer 46:2,2Ki24:7,	3489	604 BC
Nebuchadnezzar k Babylon, 4th Jehk.	Jer 25:1	3489-	604- BC
Jehoiakim a vassal 3y	2 Ki 24:1	3492-3495	601-598 BC
Babylonians defeat Jehoiakim	2 Ki 24:1-4	3495	598 BC
1st stage of exile of Judah, 7th Nebc.	Jer 52:28,25:1	3495	598 BC
Jehoiakim dep. to Babylon	2 Ch 36:6	3495	598 BC
Jehoiachin king Judah 3m10d	2 Ki 24:8,2Ch36:9	3495-3496	598-597 BC

Jehoiachin & Ezekiel dep. to Babylon	Eze 1:2,2Ki24:12	3496	597 BC
2nd stage of exile of Judah, 8th Nebc.	2 Ki 24:12	3496	597 BC
Zedekiah k Judah 10y4m[11y]	Jer 1:3,2Ki24:18,2Ch36:11	3496-3506	597-587 BC
Neb. besieges Jer, 16th Neb/9th Zedk.	2 Ki 25:1, Jer 39:1,52:4	3504	589 BC
Zedekiah dep. to Babylon	Jer 39:5-7,52:5-11	3506	587 BC
The fall of Jerusalem	Jer 52:5-7,2Ki25:2-4	3506	587 BC
3rd stage of exile Judah: Jer, 18th Nebc.	Jer 52:29,5,39:2,2Ki25:2	3506	587 BC
1st Temple of LORD burned down	**Jer 52:13,2Ki25:8-9**	**3507**	**586 BC**
The Covenant of the Temple ended			
NEBUCHADNEZZAR	***Jer 52:13, 2 Ki 25:8-9*** ***420y***	***3507***	***586 BC***

The destruction of the Temple in Jerusalem by Babylonian King Nebuchadnezzar is an extraordinary event for the Bible Chronology: **This event is the intersection between Bible History and World History**: This event is the only one that is TRULY IDENTIFIABLE by the World History and its AD/BC calendar through other sources than the Bible itself. This event takes place **in the year 586 BC**. The event is identifiable through non-Biblical sources. These are the ancient Babylonian and Assyrian calendars.

The Calendar from the Covenant of the End of Temple through Nebuchadnezzar to the Inauguration of the Second Temple by Zerubbabel

This calendar is determined by Daniel 9:2. The multiple of ten number 70 between End of the Temple of Solomon until the Inauguration of the 2nd Temple by Zerubbabel

1st Temple of LORD burned down	**Jer 52:13,2Ki25:8-9**	**3507**	**586 BC**
The fall of Egypt, 12th Eze.	Eze 32:1,17	3507	586 BC
4th stage of exile J: remnant. peop.deported	Jer 52:30,39:8-10,2Ki25:11	3511	582 BC
Evil-Merodach king Babylon	2 Ki 25:27	3532	558 BC
Jehoiachin released, 37th Jehc.	2 Ki 25:27	3532	561 BC
Persia subdues Babylon	Jer 25:11-12,2Ch36:22	3555	538 BC
The decree of Cyrus	2 Ch 36:22, Ezr1:1, Isa44:24-45:13	3555	538 BC
The 2nd Temple of the LORD started	Ezr 1:2	3555	538 BC
Daniel serves the Babylonians 70y	Da 1:1,21,2Ch36:22, Ezr1:1,	3488-3558	605-535 BC
The building of the Temple restarted	Ezr 4:24	3573	520 BC
70y withhold mercy	Zec 1:12	3503-3573	590-520 BC
The decree of Darius	Hag 1:1, Ezr5:1-6:15	3573	520 BC
The desolation of Judah 70y	Zec 1:12	3505-3575	588-518 BC
Judah brought back aft 70y, 4th Darius	Jer 29:10, Zec7:1,5	3575	518 BC
Judah to serve Babylon 70y	Jer 25:11-12, 29:10, 2 Ch36:21 Da 9:2, Zec 7:1,5, Ezr 6:16	3507-3577	586-516 BC
The 2nd Temple of the LORD completed	Da 9:2 ***70y***	3577	516 BC
The Covenant of the 2nd Temple			
Temple dedicated ZERUBBABEL	Da 9:2 ***=70y***	3577	516 BC

The Hebrew-to-Greek Old Testament Bible Translation in Alexandria, Egypt 3813 280 BC

The Septuagint translation of the Old Testament from Hebrew into Greek under Greek Alexandrian King Ptolemy II (308-246 BC) in Alexandria, Egypt marks the first step of the fulfillment of the prophecy of Genesis 9:27 – "let

Japheth dwell in the tents of Shem". The statement ""let Japheth dwell in the tents of Shem" means that spiritually the descendants of Japheth will inherit the faith of Shem which is the faith of Israel. A further milestone of this prophecy occurs with the invention of the printing press and the translation and distribution of the Bible into the Indo-European languages during the Reformation Age.

The Salvation week thrgh Jesus	Jn 12:1, 19:31, Mk 15:42, Lk 23:54	4117	25AD
The son of God, savior of all	Ge 1:1-2:25, Lk 3:38, Ac 3:15		
The 77 generations from Father to Son	Lk 3:23-38, Jude:14	0-4117	4093BC-25AD
The Covenant of Eternal Life, JESUS	Mt 26:28, Mk 14:24, Lk 22:20	4117	25AD
	1 Co 11:25, 2 Co 3:6, Heb 7:22		
The Covenant of Crucifixion	Isa 28:18, 59:21, Jer 31:31-34,		
The crucifixion of Jesus foretold	Da 9:24-26		
70 'sevens' + 7 'sevens' + 62 'sevens'	**Da 9:24-26** **= *540y***	**3577**	**25AD**
(70 x 7y) + (7x7y) + (1y + 10 weeks) = 540y			

The calendar from the Inauguration of the 2nd Temple to the Covenant of Eternal Life of Yeshua / Jesus

This Calendar is defined by the Prophecy of Daniel 9:24-26: (70 x 7y) + (7x7y) + (1y10w) = 540 years.
The correct interpretation is provided in Leviticus 25+27.
The Year of Jubilee = The Year of Cancelling Debts = the 50th year.
The 50th year is the year after the 7 'Sevens.' ***This is the year of the 62 'Sevens.'***
The fact that there is no year zero makes it challenging to determine the actual year of Yeshua/Jesus' birth and crucifixion year.

Count off seven sabbath years—seven times seven years—so that the seven sabbath years amount to a period of forty-nine years. Then have the trumpet sounded everywhere on the tenth day of the seventh month; on the Day of Atonement, sound the trumpet throughout your land. Consecrate the fiftieth year and proclaim liberty throughout the land to all its inhabitants. It shall be a jubilee for you; each of you will return to your family property and your clan. The fiftieth year shall be a jubilee for you; do not sow and do not reap what grows of itself or harvest the untended vines. For it is a jubilee and is to be holy for you; eat only what is taken directly from the fields. "'In this Year of Jubilee, everyone is to return to their property. Leviticus 25:8-13 (NIV)

THE NEW TESTAMENT: (Matthew-Revelation)

Note: The World historical calendar does NOT have the Year 0. This...3BC,2BC,1BC, AD1, AD2, AD3 ....is the World calendar's correct series of year counting.

The Time of the Earthly Life of Jesus I (a)			
Augustus/Octavian king of Roman Emp	Lk 2:1	4062-4107	AD 31 BC-14
Tiberius adopted as son & successor by Augustus		4097	AD 4
Augustus, king of Roman Empire, dies		4107	AD 14
Tiberius king of Roman Empire	Lk 3:1	4097-4130	4-14-37 AD

THE DATING OF JESUS LIFE ON EARTH - THE BIRTH MONTH OF JESUS

THE CALCULATION BASED ON ZECHARIAH'S TEMPLE SERVICE IN THE ORDER OF ABIJAH on DUTY IN 11 BC

The birth months of Jesus are determined via the Jewish calendar and Luke 1. The priest Zechariah, father of John the Baptist was in the order of Abijah on duty in the spring of 11 BC. From spring, add nine-month, and you reach winter 11/10 BC as the birth season of John the Baptist. From winter 11/10 BC, add nine months, and you get 10 BC fall (August-October) as **the birth season of Jesus**.

THE DATING OF JESUS LIFE ON EARTH - THE FIRST CALCULATION

THE CALCULATION BASED ON THE ERRONEOUS INTERPRETATION OF LUKE 3:1

"In the fifteenth year of the reign of Tiberius Caesar" =?

The first calculation of the dating of Jesus' life on earth goes to the issue's root. The question of why our World Calendar is NOT in line with the actual birth year of Jesus. Why is our World Calendar not matching with the birth year NOR with the death year of Jesus? Something went wrong with the calculation. What was it?

In around 525 AD, the Scythian Monk Dionysius Exiguus gets commissioned by Pope John I to determine a calendar for Easter. Dionysius created a new year count for the Roman State-Church in "anno domini Jesus." Dionysius names and numbers the year of Christ's birth, the year "1 AD".

What was the root cause of the erroneous dating of Jesus' birth year? We understand it was the false understanding of Luke 3:1.

The root of the error lies in the 10 years between Emperor Augustus's adoption of Tiberius and the year of Emperor Augustus's death. Augustus adopted Tiberius in 4 AD. Augustus dies in 14 AD, 10 years later.

We understand the author of the "anno domini Jesus" year count calculated ERRONEOUSLY the "fifteenth year of the reign of Tiberius Caesar" from the "death year" of Augustus. That is why he places the birth year of Jesus precisely ten years in the future or after the actual birth year of Christ Jesus.

Luke, the author of the Gospel of Luke, calculated "the fifteenth year of the reign of Tiberius Caesar" from the ADOPTION YEAR OF TIBERIUS (4 AD) and NOT from the DEATH YEAR OF AUGUSTUS (14 AD).

Hence, the actual birth year of Jesus is ten years back from the year 1 AD. Since the historical World Calendar is based on Roman numerology, there is no year zero. The birth year of Jesus is in the fall (Aug-Oct) of the year 10 BC. Jesus lived on earth 33.5 years. He dies on the cross in spring AD 25. It is the first calculation of Jesus' birth and death year based on ROMAN WORLD HISTORY (THIS IS THE EXTRA-BIBLICAL CALCULATION)!

THE DATING OF JESUS LIFE ON EARTH - THE SECOND CALCULATION
THE CALCULATION BASED ON THE CORRECT INTERPRETATION OF LUKE 3:1

Hence, the fifteenth year of Tiberius Caesar's reign = 4 AD + 14 years = 18 AD (Luke 3:1). The ministry of John the Baptist starts in 18 AD spring, 3.5 years before Jesus starts his 3.5-year ministry in 21 AD fall. Jesus' ministry lasted 3.5 years until the crucifixion in 25 AD spring. Jesus died on the cross at age 33.5. The birth year of Jesus is 10 BC fall.

THE DATING OF JESUS LIFE ON EARTH - THE THIRD CALCULATION
THE CALCULATION BASED ON MATTHEW 2:16

"Then Herod, when he saw that he was deceived by the wise men, was exceedingly angry; and he sent forth and put to death all the male children who were in Bethlehem and in all its districts, from two years old and under, according to the time which he had determined from the wise men." (NKJV, Matthew 2:16)

Matthew's story is a Deja-vue compared to Moses' story in the Book of Exodus chapter 1-2 and Abraham's account in the Book of Jasher Chapter 8-11[2]. Astrologers see signs in heaven that a future Messiah or king has been born. They speak with the reigning king about the matter. The reigning king acts out of fear and gives orders to kill the newly born child. The same story happens with Abraham, Moses, and Jesus. In all three cases, a king tries to kill a baby whom the king believes could threaten their earthly power and the Royal House.

The "Star of Bethlehem" has been deemed by many as Saturn and Jupiter planet's astronomical conjunction. The text of Matthew confirms this because it says, **"They see the star again...". Thus, they see the star multiple times during the same year. It** is not a shooting star; it is a star that appears and re-appears in the same year. Saturn and Jupiter's conjunction allowed the astronomers/astrologers to see it in the East, interpret it, follow it, and find it again when arriving in Israel.

"Where is He who has been born King of the Jews? For we have seen His star in the East and worship Him".... "When they heard the king, they departed; and behold, the star which they had seen in the East went before them, till it came and stood over where the young child was. When they saw the star, they rejoiced with exceeding great joy." (NKJV, Matthew 2:2, 9-10)

Hence, the "star" is not a "shooting star" or another unique appearance. This star is a mathematically calculable Stellar appearance occurring multiple times during the same year. It is precisely the conjunction of Saturn and Jupiter described by Johannes Kepler (1571-1630), a highly regarded scientist and believer in the Bible. This conjunction occurred in the astronomical year 7 BC, the historical year 8 BC: **Three times** May 29–June 8, September 26–October 6, and Dec 5–15.[3]

---

[2] The Bible refers to the Book of the Upright (Jasher) in Joshua 10:13 and 2 Samuel 1:18. The Book of Jasher at Sacred-Texts.com, printed first in Hebrew and published in 1613 in Venice, Italy by Yosèf ben Samuel. The Hebrew text was bought, translated and published by the Mormon (Non-Trinity-Non-Christian) J. H. Parry & Company and published in English in 1887. Sepir Ha Yasher, the Hebrew title of this book, means the 'Book of the Upright'. https://www.sacred-texts.com/chr/apo/jasher/3.htm

[3] German title: **Das Grosse Bibellexikon**, Vol. 3, p. 1483 (German edition of *The Illustrated Bible Lexicon*) Publisher: R. Brockhaus Verlag Wuppertal and Zurich, 1989

Roman Emperor Constantin (272-337 AD) placed Christmas as the festival of Christ's birth into December. In reality, December is the season when "Magi from the East" visit Jesus. The actual delivery of Jesus occurs in the late summer and fall (Aug-Oct), NOT in December. Jesus is already two years old when the three "Magi astronomers" visit him in the house, NOT in the manager.

"And when they had come into the house, they saw the young Child with Mary, His mother, and fell and worshiped Him. And when they had opened their treasures, they presented gifts to Him: gold, frankincense, and myrrh." (NKJV, Matthew 2:11). Saturn and Jupiter's conjunction occurs in 8 BC. Jesus is born 8 BC–2 years = 10 BC (fall).

In 1603, Johannes Kepler, a German astronomer, witnesses Saturn and Jupiter's conjunction 1610 years after 8 BC (7 BC astronomical calendar). Johannes Kepler realized the relationship of this conjunction with the account in the Gospel of Matthew. He published books about this matter.[4]

Herod the Great vassal king of Israel	Mt 2:1,19,22, Lk 1:5	4053-4089	40-4 BC
Archelaus tetrarch of Judea	Mt 2:22,	4089-4099	4 BC-6AD
Pilate governor of Judea	Lk 3:1, Mt 27:11, Lk 3:1	4111-4126	18-26-33AD
Herod Antipas tetrarch of Galilee	Mt 14:1, Mk 6:14, Lk 3:1	4089-4132	4 BC-39AD
Philip tetrarch of Iturea & Trac.	Mt 14:3, Mk 6:17, Lk 3:1	4089-4127	4 BC-34AD
Iturea Traconitis tetrarch of Ab.	Lk 3:1	4089-	4 BC- ....
The renovation (and expansion) of the 2nd Temple during 46y	Jn 2:20	4067-4113	26BC—21AD
Zechariah visited by an angel	Lk 1:5,24	4083	
The birth of John the Baptist, 6th m	Lk 1:24,26,36	4083	10 BC
The birth of Jesus in Bethlehem	Mt 2:16, Lk 1:56	4083	10 BC
The family of Joseph stays in Bethlehem 2y	Mt 2:16.	4083-4085	10-8 BC
The Time of the Earthly Life of Jesus I (b)			
The eastern Magicians recognize a significant star in the east:	Mt 2:2	4085	8 BC

Historical Calendar 8 BC, May 29 - June 8 (Astronomical Calendar: 7 BC, May 29 - June 8)

The first appearance of the conjunction of Saturn and Jupiter			
The eastern Magicians see the star again, during their travel which leads them to Israel.	Mt 2:7,16	4085	8 BC

Historical Calendar 8 BC, Sept. 26 – Oct. 6 (Astronomical Calendar: 7 BC, Sept. 26 – Oct. 6)

This is the second conjunction of Saturn And Jupiter. (Matthew 2:1-2) The eastern Magicians meet the vassal king Herod the

[4] German title: **Und die Bibel hat doch recht** (And still, the Bible is correct) Author: W. Keller, Publisher: Econ Verlag 1955, Germany

Great. They tell Herod that their astrological wisdom tells them that the future Messiah has already been born two years ago.			
The eastern Magicians see the star again, on their travel from Jerusalem to Bethlehem.	Mt 2:9	4085	8 BC

Historical Calendar 8 BC, Dec. 5 - 15. (Astronomical Calendar: 7 BC, Dec. 5–15)

This is the third and last appearance of the conjunction.

THE DATING OF JESUS LIFE ON EARTH - THE FOURTH CALCULATION

THE CALCULATION BASED ON REVELATION 12:6

The duration of Joseph, Mary, and Jesus escape and stay in Egypt–3.5 years: After the birth of Jesus in the fall of 10 BC, Joseph and his family stayed in Bethlehem and moved into a house. When the Magi from the East arrived at this house, two years later, they found Jesus NOT in the manger (as we know it from nativity displays) but as a walking toddler in a home. (Matthew 2:11). Hence, after the birth of his first son (concepted by the Holy Spirit) Joseph decides to live in Bethlehem for the time being. When Jesus was a two-year-old walking toddler on December 8, BC (third conjunction of Saturn and Jupiter), Joseph received a message in a dream to leave Israel and to flee to Egypt–as a place of refuge.

"The woman fled into the desert to a place prepared for her by God, where she might be taken care of for 1,260 days" (Revelation 12:6). This passage is several-layered. First, it prophesies Israel and the church's protection by God in the future. Second, it explains the history of the past, which is the earthly life of Jesus when he was brought to Egypt as a two-year-old baby.

Mary, the mother of Jesus (Revelation 12:6), stays with Joseph and Jesus in Egypt for 1,260 days, which is exactly 3.5 years. Herod the Great dies at the end of March / the beginning of April 4 BC (according to World history). Joseph, Mary, and Jesus travel from Egypt to Israel during spring and early summer 4 BC. Jesus is now a **5.5-year-old boy**!

The Time of the Earthly Life of Jesus II

Joseph is directed to escape with his family to Egypt. 8 BC, Dec	Mt 2:13-14	4085	8 BC
Herod lets the children kill, 8 BC, fall	Mt 2:17,18, Jer 31:15	4086	8 BC.
King Herod the Great dies ***4 BC, March/April***[5]	**Mt 2:15,20**	***4089***	***4 BC***
Archelaus tetrarch of Judea (4BC)	Mt 2:22	4089	4 BC
Joseph stays with his family 1260 days in ***Egypt 8 BC Nov/Dec. - 4 BC Apr/May (3.5 years)***	Rev 12:1-6	***4086-4090***	***8-4 BC***

[5] **The New Encyclopedia Britannica in 30 Volumes**, Knowledge in Depth, 15th Edition 1976, Herod I the Great

THE DATING OF JESUS LIFE ON EARTH - THE FIFTH CALCULATION

THE CALCULATION BASED ON DANIEL 9:24-26

The year of the crucifixion of Jesus and the Eternal Life's covenant is to be calculated following Daniel's prophecy **in Daniel 9:24-26.**

Through Assyrian and Babylonian calendars, we know that the BURNING OF THE TEMPLE built by Solomon occurred in the year **586 BC**. Exactly 70 years later, the Temple was rebuilt, and the Temple Service could be restarted through the Covenant of Zerubbabel (Haggai 1:14) **in 516 BC**.

From the year 516 BC, we need to add **540 years** into the future and arrive in the year 25 AD, the year of the Covenant of Eternal Life, and the Crucifixion of Jesus.

The prophet Daniel explicitly foretells the year of this new covenant under the Messiah. In Daniel 9:24-26, a vision has been given to Daniel:

"Seventy 'sevens' (or 'weeks') are decreed for your people and your holy city to finish (or restrain) transgression, to put an end to sin, to atone for wickedness, to bring in everlasting righteousness, to seal up vision and prophecy and to anoint the holiest (or Most Holy Place)." (Daniel 9:24, NIV)

From the Temple Inauguration by Zerubbabel in 516 BC, add 490 years (70 x 7). This is the year 26 BC. In 26 BC, Herod the Great receives a decree from Rome to rebuild Jerusalem (and the Temple).

"Know and understand this: From the time the word goes out to restore and rebuild Jerusalem until the Anointed One, the ruler, comes, there will be seven' sevens,' and sixty-two 'sevens.' It will be rebuilt with streets and a trench, but in times of trouble. After the sixty-two 'sevens,' the Anointed One will be put to death and will have nothing. (Daniel 9:25-26, NIV)

From the Roman decree to rebuild Jerusalem, add 49 years (7 x 7). These are the 49 years before the year of Jubilee mentioned in Leviticus 25:8-52 and 27:17-24, the year 24 AD, one year before the crucifixion year of Jesus.

To the year before the crucifixion year of Jesus, which is the year 24 AD, add one year (the Year of Jubilee–or the fiftieth Year) - these are the 62 'Sevens'–or one year and ten weeks, to arrive in the crucifixion's year of Jesus which is the year 25 AD.

From the year 25 AD, spring subtract 33.5 years. Jesus is born in the fall of 10 BC.

Jesus begins his ministry at age 30–Luke 3:23 in the fall of 21 AD. Jesus' ministry lasts 3.5 years from fall 21 AD to spring 25 AD.

THE DATING OF JESUS LIFE ON EARTH - THE SIXTH CALCULATION

THE CALCULATION BASED ON JOHN 2:13, 6:4, 8:2 AND 12:1

The Gospel of John provides four passages where Jesus teaches in the Temple in Jerusalem. Based on this data, we understand that Jesus' ministry begins in the fall of 21 AD after which it underwent four visits to Jerusalem during four Passover Feasts by Jesus and the disciples in **22 AD, 23 AD, 24 AD, and 25 AD. (JOHN 2:13, 6:4, 8:2 and 12:1).**

This is in line with the understanding Jesus' ministry lasted 3.5 years. The 3.5 years match with the 3.5 years of future events described in Revelation 11:3–1260 days. The two witnesses of Jesus' earthly ministry are the Son and the Holy Spirit–with Jesus' earthly life.

THE DATING OF JESUS LIFE ON EARTH - THE SEVENTH CALCULATION

THE CALCULATION BASED ON JOHN 2:20

They replied, "It has taken forty-six years to build this temple, and you are going to raise it in three days?"

Count from Zerubbabel's Covenant in 516 BC–490 years (future) to 26 BC–the Roman decree year to rebuild Jerusalem (and the Temple).

Herod the Great receives the decree to rebuild Jerusalem in the fall of 26 BC. Construction work to rebuild Jerusalem and the Temple started in 25 BC, spring.

Add 46 years from 25 BC, spring = 22 AD spring. Jesus starts his ministry in 21 AD, fall. In 22 AD spring, Jesus visits Jerusalem during the Passover Feast **for the first time**–since his baptism in the Jordan. Jesus' disciples meet a group of people who make the statement: "It has taken forty-six years to build this temple, and you are going to raise it in three days?"

After 22 AD spring, Jesus returns to Jerusalem during the Passover Feast in 23, 24, and 25 AD (JOHN 2:13, 6:4, 8:2 AND 12:1)

Joseph returns with his family back to Israel	Mt 2:19-23		
4 BC spring		4090	4 BC
Joseph moves with his family to Nazareth	Mt 2:22-23		
4 BC summer		4090	4 BC
The account of Jesus as a 12-year-old in	*Lk 2:41,42,46,51*	*4096*	*3 AD*
The Temple of the LORD, 3 AD, Jesus 12y			
The decree of the Roman Empire to	***Da 9:24***	***4067***	***26 BC***
Restore Jerusalem and the Temple of the LORD, 26 BC			
The renovation (and expansion)	***Jn 2:20***	***4068-4114***	***25BC-22AD***
of the 2nd Temple during 46y (25+22-1)			
John the Baptist starts his ministry in	***Lk 3:1, Mt3:1, Isa40:3***	***4111***	***18AD***
the 15th year of Tiberius = 18 AD			

The ministry of John the Baptist, (5y)	***Lk 3:1-19, Jn1:19-40***		***4111-4115***	***18-23AD***
Jesus arrives from Nazareth in Judea	Mt 3:13, Mk1:9		4113	21AD
The Baptism of Jesus,	***Lk 3:21, Mt3:13, Mk1:9-11***		***4113***	***21AD***
Jesus 30years old 21AD fall/autumn				
The Time of the Coming of Jesus upon the Earth/ The Word becomes Flesh John 1:14				
David king over Judah in Hebron 7y6m	2 Sam 2:11,1Ki2:11,1Ch29:26 (6m)		3043-3050	1050-1043 BC
David king over Israel in Jerus. 33y	2 Sam 5:4,5,1Ki2:11,1Ch3:4		3050-3083	1043-1010 BC
The life of David 70y	2 Sam 5:4		3013-3083	1080-1010
The Ark brought to Jerusalem	2 Sam 6:1,11,12		3050	1043 BC
The death of David	1 Ki 2:10		3083	1010 BC
Solomon king over Israel 40y	1 Ki 2:12,11:42,2Ch9:30		3083-3123	1010-970 BC
Solomon starts building the			3086	1007 BC
Temple of the LORD, 480th Exodus	1 Ki 6:1, [LXX: 1Ki6:1,440th]			
Solomon king Israel,4th Solomon	2 Ch 3:2		3086	1007 BC
The Covenant of the Temple, SOLOMON	1 Ki 6:1	***480y***	3087	1006 BC
(Temple under construction one year)				
The 1st Temple of the LORD, NEBUCHD. destroyed	Jer 52:13, 2 Ki 25:8-9	***420y***	3507	586 BC
The 1st Temple of the LORD, NEBUCHD. destroyed	Jer 52:13, 2 Ki 25:8-9	420y	3507	586 BC
The fall of Egypt 12th Eze.	Eze 32:1,17		3507	586 BC
4th stage of exile Judah:rem.peop.dep.ted	Jer 52:30,39:8-10,2Ki25:11		3511	582 BC
Evil-Merodach king Babylon	2 Ki 25:27		3532	561 BC
Jehoiachin released 37th Jehc.	2 Ki 25:27		3532	561 BC
Persia subdues Babylon	Jer 25:11-12,2Ch36:22		3555	538 BC
The decree of Cyrus	2 Ch 36:22,Ezr1:1,Isa44:24-45:13		3555	538 BC
The 2nd temple of the LORD started	Ezr 1:2		3555	538 BC
Daniel serves the Babylonians 70y	Da 1:1,21,2Ch36:22,Ezr1:1,		3488-3558	605-535 BC
The building of the temple restarted	Ezr 4:24		3573	520 BC
70y withhold mercy	Zec 1:12		3503-3573	590-520 BC
The decree of Darius	Hag 1:1,Ezr5:1-6:15		3573	520 BC
The desolation of Judah 70y	Zec 1:12		3505-3575	588-518 BC
Judah brought back aft 70y 4th Darius	Jer 29:10,Zec7:1,5		3575	518 BC
Judah to serve Babylon 70y	Jer 25:11-12,29:10,2Ch36:21,			
	Da 9:2,Zec 7:1,5		3507-3577	586-516 BC
The Covenant of the 2nd Temple Dedicated by ZERUBBABEL	Da 9:2	***70y***	3577	516 BC
The crucifixion of Jesus foretold	Da 9:24-26	***540y***	3577-4117	516BC-25AD
70 'sevens' + 7 'sevens' + 62 'sevens'				
= (70 x 7y) + (7 x 7y) + 1y = 540y	Da 9:25-26, Mt 27:35	***540y***	4117	25 AD
The 70 sevens: (490 years) 516 - 26 BC	Da 9:24	***490y***	3577-4067	516-26 BC

Event	Reference	Years	AM	Date
From the dedication of the Second Temple until the decree of Caesar given to Herod the Great to renovate the Second Temple (3rd Temple for Christ)				
The 7 sevens: (49 years) 26 BC - 24AD	Da 9:25	***49y***	4067-4116	26 BC-24AD
From the Roman decree to renovate the Second Temple until the beginning of the last year of Jesus‘ ministry.				
The 62 sevens: (1year + 10 weeks) 24AD - 25AD	Da 9:26	***1y***	4116-4117	24AD-25
From the beginning of the last year of Jesus' ministry until his crucifixion.				
The 2nd (pre-3rd) Temple of the LORD being renovated during 46 years 26 BC - 21AD (46=26+21-1)	Jn 2:20	***46y***	4067-4113	26 BC-21AD
The ministry of Jesus: 3.5 years		***3.5y***	4113-4117	21AD-25AD

I The Time of the Earthly Life of Jesus III
II The Time of the ministry of Jesus: 3.5 years Revelation 11:3
III The First Year of the ministry of Jesus John2:13,23,4:45

Event	Reference	Years	AM	Date
Jesus comes from Nazareth to Judea	Mt 3:13, Mk1:9		4113	21AD
The Baptism of Jesus,				
Jesus 30y, fall/autumn 21AD	***Lk 3:21, Mt 3:13, Mk 1:9-11***	***0y***	***4113***	***21AD***
The ministry of Jesus: 3.5 years	***Rev 11:3***		***4113-4117***	***21AD-25AD***
21AD autumn - 25AD spring	.		4113	21AD
Jesus starts his ministry	Lk 3:23		4113	21AD
Jesus in the desert of Judea,40d	Lk 4:2, Mt4:1, Mk1:12		4113	21AD
Jesus in Bethany, Judea	***Jn 1:28***		***4114***	***22AD***
2 disciples: [Simon-Peter+Andrew], Judea	Jn 1:35-42		4114	22AD
Jesus returns to Galilee	Jn 1:43		4114	22AD
Jesus in Cana, at a wedding,1st wonder	Jn 2:1,11		4114	22AD
Jesus in Capernaum	Jn 2:12,		4114	22AD
The New Temple by Herod the Great	***Jn 2:20***		***4114***	***22AD***
completed in its renovation: 21AD	.		*4114*	*22AD*
The 1st Passover, spring 22 AD	***Jn 2:13,23,4:45***	***+ 0.5y = 0.5y***	***4114***	***22AD***

The second year of the ministry of Jesus John2:13,23,4:45

Event	Reference	Years	AM	Date
The 1st Passover, Jesus in Jerusalem	***Jn 2:13,23,4:45***	***+ 0.5y = 0.5y***	***4114***	***22AD***
Jesus teaches after his first ministry-year during the Passover Feast in the temple	John 2:13,23, 4:45			

of the LORD in Jerusalem		4114	22AD
The temple of the LORD cleansed	Jn 2:14	4114	22AD
Jesus in Jerusalem, spring 22 AD	***Lk 4:9, Mt4:5***	***4114***	***22AD***
Jesus talking with Nicodemus	Jn 3:1-21	4114	22AD
Babysitting in Judea	Jn 3:22	4114	22AD
John the Baptist at Aenon near Salim	Jn 3:23	4114	22AD
John the Baptist taken captive	Mt 4:12,14:3, Mk1:14,6:17, Lk3:19-20	4114	22AD
Jesus moves to Galilee	Jn 4:3,43, Lk 4:14, Mt4:12	4114	22AD
Jesus in Sychar, talking with Sam.wom.	Jn 4:5	4114	22AD
disciples rejoin Jesus in Sychar	Jn 4:5,27	4114	22AD
The Samaritans of Sychar belief	Jn 4:5,39-42	4114	22AD
Jesus in Cana	Jn 4:46	4114	22AD
Jesus in Capernaum,2nd wonder	Jn 4:46,54	4114	22AD
Jesus in Nazareth	Lk 4:16,23	4114	22AD
Jesus in Capernaum, driving out E.S.	Lk 4:31, Mk1:21	4114	22AD
At the Sea of Galilee near Capernaum	Lk 5:1, Mt4:18, Mk1:16,21	4114	22AD
2 disciples: Simon/Peter, Andrew, at S	Lk 5:1-11, Mt4:18, Mk1:16	4114	22AD
2 disciples: James, John, at the Sea of G.	Mk 1:19, Mt 4:21	4114	22AD
The healing of many in Capernaum	Mk 1:29-34,21	4114	22AD
Jesus prays on a mountain, at Caper.	Mk 1:35-37,21	4114	22AD
The ministry in Galilee	Mk 1:39, Mt4:23	4114	22AD
Jesus prays on a mountain, at Caper.	Lk 6:12, Mk3:13	4114	22AD
Leaving Capernaum	Mk 1:38	4114	22AD
Healing a man with leprosy	Mk 1:40-45, Mt8:1-4, Lk5:12-16	4114	22AD
Jesus in Capernaum, healing a papal.	Mk 2:1, Mt9:1	4114	22AD
At the sea of Galilee	Mk 2:13, Mt9:1	4114	22AD
1 disciple: Levi/Matthew, at Sea of G.	Mk 2:13, Lk5:27, Mt9:9	4114	22AD
The 12 disciples chosen	Lk 6:13, Mk3:13, Mt10:1	4114	22AD
The Sermon on the mount	Lk 6:17-49, Mt5:1-7:29	4114	22AD
Jesus in Capernaum, faith centurion	Mt 8:5	4114	22AD
The question of John's disciples	Mk 2:18, Mt9:14	4114	22AD
The question of John the Baptist	Lk 7:18, Mt11:2	4114	22AD
Jesus in Capernaum, centurion's faith	Lk 7:1, Mt8:5	4114	22AD
Jesus in Nain, son of widow, raised	Lk 7:11	4114	22AD
Jesus in Gedarenes, healing Dem. Pos.	Mk 5:14, Lk 8:26, Mt8:28	4114	22AD
Word for Decapolis	***Mk 5:20***	***4115***	***23AD***
John the Baptist murdered, 23AD	*Mk 6:27,9:13, Mt 14:10, Mt17:12-13*	**4115**	***23AD***
The 12 disciples sent	Mk 6:7, Lk9:1, Mt10:1,5	4115	23AD
Herod hears about Jesus	Mk 6:14, Mt14:1	4115	23AD
The disciples rejoin Jesus	Lk 9:10	4115	23AD
Jesus in Bethsaida	Lk 9:10, Mk8:22	4115	23AD
The 5000 fed	Jn 6:4,10, Mk 6:30, Mt14:30, Lk9:10,14	4115	23AD

Jesus walking on the Sea	Jn 6:16-17,25, Mk 6:48, Mt14:25	4115	23AD
Jesus in Capernaum/Gennesaret	Mk 6:53, Mt14:34	4115	23AD
The third year of the ministry of Jesus John2:13,23,4:45			
The 2nd Passover, spring 23 AD	***Jn 6:4,7:1 0.5y+1y = 1.5y***	***4115***	***23AD***
Jesus teaches after his second ministry-year during the Passover Feast in the Temple of the LORD in Jerusalem,	*Jn 6:4*	4115	23AD
The Pharisees come from Jerusalem	Mk 7:1, Mt15:1,12:2,24,38	4115	23AD
The ministry in Phoenicia	Mk 7:24	4115	23AD
Jesus in Tyre and Sidon	Mk 7:24, Mt15:21,11:21	4115	23AD
Jesus in Capernaum	Mt 12:46,13:1	4115	23AD
Sidon & Tyre compared to Krasin & B.	Mt 11:21	4115	23AD
The ministry in Decapolis	Mk 7:31	4115	23AD
At the sea of Galilee	Mt 15:29	4115	23AD
The 4000 fed	Mt 15:29, Mk8:1,9,	4115	23AD
Jesus in the region of Samantha	Mk 8:10	4115	23AD
Jesus in Magadan	Mt 15:39	4115	23AD
Jesus in Jerusalem, feast of Tabern., Fall	***Jn 7:2,10,14,25,37***	***4115***	***23AD***
On the Mount of Olives	***Jn 8:1 0.5y+1y+1y = 2.5y***	***4116***	***24AD***
In Temple Court 3rd Passover, 24 AD	***Jn 8:2,59 0.5y+1y+1y = 2.5y***	***4116***	***24AD***
The final ministry of Jesus starts: the 62 Sevens (62 weeks)	*Da 9:25*	*4116*	*24AD*
The Pharisees and Sadducees visit J.	Mt 16;1,5, Mk 7:11	4116	24AD
On the lake of Galilee	Mt 16:5, Mk7:14	4116	24AD
Jesus in Bethsaida	Mk 8:22	4116	24AD
Jesus in Caesarea Philippi	Mt 16:13, Mk8:27	4116	24AD
Jesus on the Mount of Transfiguration	Mt 17:1, Mk9:2, Lk 9:28	4116	24AD
At the Mount of T, boy healed from E.S.	Mt 17:15, Mk9:17, Lk9:37	4116	24AD
Jesus in Capernaum, disc. on the great.	Mt 18:1, Mk9:33, Lk9:46	4116	24AD
Jesus moves to Judea, acr. the Jordan	Mt 19:1, Mk10:1	4116	24AD
Jesus in Jerusalem, feast of Dedic., Fall	***Jn 10:22-23***	***4116***	***24AD***
At the Jordan, Judea	Jn 10:40	4116	24AD
In Bethany, Judea (Death of Lazarus)	***Jn 11:1,18***	***4117***	***25AD***
Jesus moves towards Jerusalem	Lk 9:51	4117	25AD
Jesus sends 72	Lk 10:1	4117	25AD
Jesus moves towards Jerusalem	Lk 13:22	4117	25AD
Jesus in Jericho	Mt 20:29, Mk10:46	4117	25AD
Jesus at Bethpage	Mt 21:1, Mk11:1	4117	25AD
The 4th Passover, the crucifixion	***Jn 12:1,12,13:1, Mt26:2 3.5y***	***4117***	***25AD***

I The Time of the final week of Jesus' ministry/ The Creation-Salvation Week/
II The Time of the Final Year of Jesus' ministry, the 62 Sevens Daniel 9:25

III The Crucifixion of Jesus, The HOLY TRINITY set free
IV The Time of the Death of Jesus; The Time of Resurrection and Ascension.

The ministry of Jesus: 3.5 years

21AD autumn - 25AD spring	***Rev 11:3***	***0.5y+1y+1y+1y = 3.5y***	***4113-4117***	***21-25AD***
The 4th Passover, the crucifixion	***Jn 12:1,12,13:1, Mt26:2***	***3.5y***	***4117***	***25AD***
Jesus buried for three days	Mt 12:40		4117	25AD
The 62 Sevens are completed: Jesus	Jn 12:1Da 9:25-26		4117	25AD
teaches during the Preparation Week	Jn 12:1Da 9:25-26		4117	25AD
in the Temple of the LORD,	Jn 12:1Da 9:25-26		4117	25AD
Jesus on the Mount of Olives	Mt 21:1, Mk11:1		4117	25AD
Jesus moves to Jerusalem	Mt 20:17, Mk11:1,12		4117	25AD
Jesus in Jerusalem the next day again	Mt 21:18, Mk11:20		4117	25AD
Jesus in the temple of the LORD	Mt 21:23,24:1, Lk 19:45		4117	25AD
The LORD's supper/				
The covenant of Life, JESUS	Mt 26:17		4117	25AD
Jesus in the Kidron Valley/Gethsemane	Jn 18:1, Mt 26:36		4117	25AD
Jesus arrested	Jn 18:1		4117	25AD
The decision to kill Jesus	Mt 27:1		4117	25AD
Jesus before Pilate	Mt 27:11,13		4117	25AD
The Day before Sabbath, Day of Preparation			4117	25AD
The 6th Day, The Day of Man's Creation			4117	25AD
Jesus in Jerusalem in the Temple	Mt 27:35, Da 9:25-26		4117	25AD
The 4th Passover, the crucifixion	***Jn 12:1,12,13:1, Mt26:2***	***3.5y***	***4117***	***25AD***
Jesus buried for three days	Mt 12:40		4117	25AD
The salvation week thr JESUS, 1 week	Mt, Mk, Lk, Jn		4117	25AD
The last week of the earthly life of Jesus	Mt, Mk, Lk, Jn		4117	25AD
The creation week thrgh JESUS, 1 week	Ge 1:26, 2:7, Ex 20:11,31:17	***0y***	0	4093 BC
The salvation week thr JESUS, 1 week	Mt, Mk, Lk, Jn	***4117y***	4117	25AD
The 3rd Passover, the crucifixion	Jn 12:1,12,13:1, Mt26:2	***1y***	4117	25AD
The ministry of Jesus: 3.5 years				
21AD autumn - 25AD spring	Rev 11:3	***3.5y***	4113-4117	20-25AD
The crucifixion of Jesus, JESUS	Mk 15:42, Jn 19:18-42		4117	25AD
The crucifixion of Jesus foretold	Da 9:24-26			
70 'sevens' + 7 'sevens' + 62 'sevens'				
= (70 x 7y) + (7 x 7y) + 1y (+ 10 w) = 540y		***540y***	3577-4117	516 BC-25AD
The son of God, savior of all	Ge 1:1-2:25, Lk 3:38, Ac 3:15			
The 77 generations from Father to Son	Lk 3:23-38, Jude:14		0-4117	4093BC-25AD
The Salvation week thrgh Jesus	Jn 12:1, 19:31, Mk 15:42, Lk 23:54		4117	25AD
The Covenant of Crucifixion/the Blood/	Isa 28:18, 59:21, Jer 31:31-34,			
The Covenant of Eternal Life, JESUS	Mt 26:28, Mk 14:24, Lk 22:20		4117	25AD
The resurrection of Jesus	Jn 20:1,14,19		4117	25AD

The 1st appearance to Mary Magdalene	Jn 20:1,14	4117	25AD
The 2nd appearance to some disciples	Jn 20:19	4117	25AD
The 3rd appearance to Thomas,1w later	Jn 20:26	4117	25AD
The appearance at the Sea of Tiberias	Jn 21:1,4	4117	25AD
The 3rd ministry in Galilee			
The 4th ministry in Galilee	Mt 15:39,17:24		
The time until the preaching of Paul 70 sevens, 490y	Da 9:24	3635-4125	516-26 BC
Caligula king of Roman Empire		4130-4134	37AD-41
Claudius king of Roman Empire		4134-4147	41AD-54
Nero king of Roman Empire		4147-4161	54AD-68
Vespasian king of Roman Empire		4162-4172	69AD-79

Note: The World historical calendar does NOT have the Year 0. This...3BC,2BC,1BC, 1AD, 2AD, 3AD....is the World calendar's correct series of year counting.

Today's Date: Year 2021 AD		6113	2021AD
From Creation to Salvation	**4117 years**	0-4117	4093BC – 25AD
Since Creation, The age of the earth (in 2021AD)	**6113 years**	0-6104	4093BC – 2021AD
Since Christ's Earthly Birth Date (in 2021AD)	**2030 years**	4083-6113	10BC – 2021AD
Since Crucifixion (in 2021AD)	**1996 years**	4117-6113	25AD – 2021AD

Time References in the Book of Acts:

Barnabas, Saul teaches in Antioch 1y	Acts 11:26	No Bible figures available
Paul teaches in Corinth 1 y, 6m	Acts 18:11	No Bible figures available
Paul teaches in Ephesus 2y	Acts 19:10	No Bible figures available
Paul teaches in Asia 3 years	Acts 20:31	No Bible figures available
Two years between Felix & Porcius Festus	Acts 24:27	No Bible figures available
Paul stays two years in Rome	Acts 28:30	No Bible figures available
Corinthian church started 1 year ago	2 Cor 8:10	No Bible figures available
Achaia was ready 1 year ago	2 Cor 9:2	No Bible figures available
Paul's ministry last min 14y	2 Cor 12:2	No Bible figures available
Paul meets Peter in Jerusalem after 3y.	Gal 1:18	No Bible figures available
Paul with Barnabas in Jerusalem after 14y.	Gal 2:1	No Bible figures available

<u>The End Times:</u>

The Covenant of the Crucifixion of Christ	4117	25 AD
The Rapture of the Church	4117+X	25+X AD
The 7-Year Tribulation Time	4117+X+7	25+X+7 AD
The 1000-Year Kingdom, beginning	4117+X+7	25+X+7 AD
The 1000-Year Kingdom, end	4117+X+7+1000	25+X+7+1000 AD
The End of the Earth	4117+X+7+1000	25+X+7+1000 AD

www.ingramcontent.com/pod-product-compliance
Lightning Source LLC
LaVergne TN
LVHW070202110826
845147LV00002B/471

* 9 7 8 1 7 3 6 6 2 7 0 1 3 *